T0402111

BIRKHÄUSER

Progress in Inflammation Research

Forthcoming titles:

New Therapeutic Targets in Rheumatoid Arthritis, P.-P. Tak (Editor), 2009
Inflammatory Cardiomyopathy (DCM) – Pathogenesis and Therapy, H.-P. Schultheiß, M. Noutsias (Editors), 2009
Th 17 Cells: Role in Inflammation and Autoimmune Disease, B. Ryffel, F. Di Padova (Editors), 2009
Occupational Asthma, T. Sigsgaard, D. Heederick (Editors), 2009
Nuclear Receptors and Inflammation, G.Z. Feuerstein, L.P. Freedman, C.K. Glass (Editors), 2009

(Already published titles see last page.)

Microarrays in Inflammation

Andreas Bosio
Bernhard Gerstmayer

Editors

Birkhäuser
Basel · Boston · Berlin

Editors

Andreas Bosio
Research & Development
Miltenyi Biotec GmbH
Friedrich-Ebert-Straße 68
51429 Bergisch Gladbach
Germany

Bernhard Gerstmayer
Miltenyi Biotec GmbH
Friedrich-Ebert-Straße 68
51429 Bergisch Gladbach
Germany

Library of Congress Control Number: 2008936304

Bibliographic information published by Die Deutsche Bibliothek
Die Deutsche Bibliothek lists this publication in the Deutsche Nationalbibliografie;
detailed bibliographic data is available in the internet at http://dnb.ddb.de

ISBN 978-3-7643-8333-6 Birkhäuser Verlag AG, Basel – Boston – Berlin

Basel · Boston · Berlin
P.O. Box 133, CH-4010 Basel, Switzerland
Part of Springer Science+Business Media
Printed on acid-free paper produced from chlorine-free pulp. TCF ∞
Cover design: Markus Etterich, Basel
Cover illustration: with friendly permission of Bernhard Gerstmayer
Printed in Germany
ISBN 978-3-7643-8333-6

e-ISBN 978-3-7643-8334-3

9 8 7 6 5 4 3 2 1

www.birkhauser.ch

Contents

List of contributors

Victor Appay, Immunologie Cellulaire et Tissulaire, INSERM U543, Faculté de Médecine, Hôpital Pitié-Salpêtrière, 91 Bd de l'Hôpital, 75634 Paris Cedex 13, France; e-mail: victor.appay@chups.jussieu.fr

Kazuhiko Arima, Division of Medical Biochemistry, Department of Biomolecular Sciences, Saga Medical School, Saga, 849-8501, Japan; e-mail: arimaka@cc.saga-u.ac.jp

Sergio E. Baranzini, Department of Neurology, School of Medicine, University of California, San Francisco, Medical Sciences Building S-256, 513 Parnassus Ave., San Francisco, CA 94143, USA; e-mail: sebaran@cgl.ucsf.edu

Andreas Bosio, Research & Development, Miltenyi Biotec GmbH, Friedrich-Ebert-Straße 68, 51429 Bergisch Gladbach; email: andreas.bosio@miltenyibiotec.de

Verena Brune, Institut für Zellbiologie (Tumorforschung), Universitätsklinikum, Universität Duisburg-Essen, Virchowstr. 173, 45122 Essen, Germany

Gerd Burmester, Department of Rheumatology and Clinical Immunology, Charité – Universitätsmedizin Berlin, Charitéplatz 1, 10117 Berlin, Germany; e-mail: gerd.burmester@charite.de

Roland Frötschl, Federal Institute for Drugs and Medical Devices (BfArM), Kurt-Georg-Kiesinger-Allee 3, 53175 Bonn, Germany; e-mail: r.froetschl@bfarm.de

Bernhard Gerstmayer, Miltenyi Biotec GmbH, Friedrich-Ebert-Straße 68, 51429 Bergisch Gladbach, Germany; e-mail: bernhard.gerstmayer@miltenyibiotec.de

Stephen D. Ginsberg, Center for Dementia Research, Nathan Kline Institute, New York University School of Medicine, 140 Old Orangeburg Road, Orangeburg, NY 10962, USA; e-mail: ginsberg@nki.rfmh.org

Andreas Grützkau, German Arthritis Research Center (DRFZ), Charitéplatz 1, 10117 Berlin, Germany; e-mail: Gruetzkau@drfz.de

Claudia H. Hartmann, Department of Pathology, Division of Oncogenomics, University of Regensburg, Franz-Josef-Strauss-Allee 11, 93053 Regensburg, Germany; e-mail: Claudia.Hartmann@klinik.uni-regensburg.de

Thomas Häupl, Department of Rheumatology and Clinical Immunology, Charité – Universitätsmedizin Berlin, Charitéplatz 1, 10117 Berlin, Germany; e-mail: thomas.haeupl@charite.de

Damir Herman, Myeloma Institute for Research and Therapy, University of Arkansas for Medical Sciences, 4301 West Markham, Slot #776, Little Rock, AR 72205, USA; e-mail: dherman@uams.edu

Olaf Holtkötter, Phenion GmbH & Co. KG, Merowingerplatz 1a, 40225 Düsseldorf, Germany

Kenji Izuhara, Division of Medical Biochemistry, Department of Biomolecular Sciences, Saga Medical School, Saga, 849-8501, Japan; e-mail: kizuhara@cc.saga-u.ac.up

Sachiko Kanaji, Division of Medical Biochemistry, Department of Biomolecular Sciences, Saga Medical School, Saga, 849-8501, Japan

Peter Kasper, Federal Institute for Drugs and Medical Devices (BfArM), Kurt-Georg-Kiesinger-Allee 3, 53175 Bonn, Germany

Christoph A. Klein, Department of Pathology, Division of Oncogenomics, University of Regensburg, Franz-Josef-Strauss-Allee 11, 93053 Regensburg, Germany; e-mail: Christoph.Klein@klinik.uni-regensburg.de

Ralf Küppers, Institut für Zellbiologie (Tumorforschung), Universitätsklinikum, Universität Duisburg-Essen, Virchowstr. 173, 45122 Essen, Germany; e-mail: ralf.kueppers@uk-essen.de

Martin Larsen, Immunologie Cellulaire et Tissulaire, INSERM U543, Faculté de Médecine, Hôpital Pitié-Salpêtrière, 91 Bd de l'Hôpital, 75634 Paris Cedex 13, France; e-mail: larsen@chups.jussieu.fr

Shoichiro Ohta, Division of Medical Biochemistry, Department of Biomolecular Sciences, Saga Medical School, Saga, 849-8501, Japan; e-mail: ohtasho@cc.saga-u.ac.jp

Dirk Petersohn, Phenion GmbH & Co. KG, Merowingerplatz 1a, 40225 Düsseldorf, Germany; e-mail: dirk.petersohn@henkel.com

Andreas Radbruch, German Arthritis Research Center (DRFZ), Charitéplatz 1, 10117 Berlin, Germany; e-mail: radbruch@drfz.de

Birgit Sawitzki, Institute of Medical Immunology, Charité – Universitätsmedizin Berlin, Charitéplatz 1, 10117 Berlin, Germany; e-mail: birgit.sawitzki@charite.de

Jürgen Schmitz, Miltenyi Biotec GmbH, Friedrich-Ebert-Straße 68, 51429 Bergisch Gladbach, Germany; e-mail: juergens@miltenyibiotec.de

Hiroshi Shiraishi, Division of Medical Biochemistry, Department of Biomolecular Sciences, Saga Medical School, Saga, 849-8501, Japan; e-mail: shiroshi@cc.saga-u.ac.jp

Karl Skriner, Department of Rheumatology and Clinical Immunology, Charité – Universitätsmedizin Berlin, Charitéplatz 1, 10117 Berlin, Germany; e-mail: karl.skriner@charite.de

Bruno Stuhlmüller, Department of Rheumatology and Clinical Immunology, Charité – Universitätsmedizin Berlin, Charitéplatz 1, 10117 Berlin, Germany; e-mail: bruno.stuhlmueller@charite.de

Enrico Tiacci, Institut für Zellbiologie (Tumorforschung), Universitätsklinikum, Universität Duisburg-Essen, Virchowstr. 173, 45122 Essen, Germany

Nathalie Viguerie, Inserm, U858, Laboratoire de Recherches sur les Obésités, Institut de Médecine Moléculaire de Rangueil, 31432 Toulouse; Université Paul Sabatier, Institut Louis Bugnard IFR31, 31432 Toulouse; Centre Hospitalier Universitaire de Toulouse, 31059 Toulouse, France; e-mail: Nathalie.Viguerie@inserm.fr

Hans-Dieter Volk, Institute of Medical Immunology, Charité – Universitätsmedizin Berlin, Charitéplatz 1, 10117 Berlin, Germany

Noriko Yuyama, Genox Research, Inc., Tokyo, 154-0004, Japan; e-mail: n-yuyama @kirin.co.jp

Preface

Microarray technology is applied extensively in the field of inflammation research. During the last decade, novel array-based technologies have led to insights into pathological pathways such as asthma, inflammatory skin diseases, neuroinflammation, rheumatology, and other autoimmune diseases. These insights have broadened our general knowledge of inflammatory processes and have already translated into drug development and clinical application.

While microarrays have evolved technically to the highest standards, the challenges have shifted towards improvements in sample processing, including cell sorting, microdissection, RNA preservation, RNA amplification methods, and bioinformatics.

This book, therefore, captures expert knowledge and provides in-depth information on the application of microarrays to our understanding of inflammation.

The book is divided into two major sections, beginning with the peculiarities of cell sorting, RNA isolation, and amplification methods for the most prominent tissues subject to inflammation. In the second section, highly relevant research topics in inflammatory diseases using microarray based approaches are described. In addition, an update of microarrays in drug development is presented from a regulatory perspective.

We do hope that this book will allow the reader to deepen his competence in the planning, performing, analysis and critical review of genomics experiments, as well as his or her knowledge of the current status of inflammation research.

The recent advent in high throughput sequencing technologies will very likely complement classic microarray-based approaches in deciphering the complexity of transcriptomes. Still, appropriate sample processing and interpretation of huge and complex datasets for the purpose of generating biologically meaningful results remain challenging.

Finally, we are grateful to all contributors who agreed to share their expertise from very divergent disciplines, enabling us to write this book, unifying the fascinating fields of microarray technologies and inflammation research.

July 2008 Andreas Bosio and Bernhard Gerstmayer

Microarray and inflammation: an introduction

Andreas Bosio

Miltenyi Biotec GmbH, Friedrich-Ebert-Straße 68, 51429 Bergisch Gladbach, Germany

Abstract

The history of microarrays applied to inflammation research goes back to the very first DNA microarrays ever used in 1995. Since then, several thousand reports have been published analysing almost all forms of inflammation in all kinds of tissue. Moreover, besides mRNAs, different classes of biomolecules have been addressed using different microarray designs, with microRNAs now being at the forefront of microarray based inflammation research. A general description of microarrays and their handling is followed by a survey of some central aspects of microarray data acquisition, analysis, and mining. A number of mandatory prerequisites for sample preparation are outlined with respect to the extraordinary sensitivity of microarrays and the particular situation in inflammation research. It is stressed that many of the transcriptome changes induced by inappropriate sample processing are indistinguishable from the physiological inflammation processes of interest. Finally, strategies as to how to enhance sensitivity by reducing the complexity of a sample are discussed.

A brief history of microarrays in inflammation research

Microarray technology has evolved into the most valuable tool for the recognition of genomic signatures of biological conditions which can be interpreted as a precise molecular phenotype of cell cultures or tissues in a specific state. It has been almost 20 years since the first protocols for the production of microarrays were published [1–4]. At first these arrays were intended for DNA sequencing and point mutation detection, but it became soon clear that they were unsuited to displace conventional sequencing technologies. Paralleling the increasing number of characterized genes and expressed sequence tags (ESTs), another application entered the tantalizing field of DNA arrays: gene expression profiling. The first reports of massive parallel gene expression profiling using gene arrays revolutionized the understanding of how complex gene interactions could influence the state of living cells [5, 6]. From then on, microarray literature was governed by reports describing all kinds of method-

Microarrays in Inflammation, edited by Andreas Bosio and Bernhard Gerstmayer

ologies for the production of microarrays, for sample processing, target labelling, hybridisation, scanning, as well as data analysis and deposition [7]. Many labs started to produce their own arrays but only a few had access to appropriate hardware and were able to logistically handle thousands of probes. Also, the available sequence data and cDNA clone libraries were of poor quality [8]. Then, the human genome was completed and, correspondingly, the quality of cDNA sequences was significantly improved [9, 10]. Furthermore, substantial worldwide efforts were made to standardize for the generation, presentation, and exchange of microarray data [11]. These facts, together with an increasing number of companies producing microarrays on an industrial scale, ultimately raised the standard of microarray analysis to a level permitting the generation of robust biological data.

Interestingly enough, right from the beginning inflammation processes were at the very centre of interest for those applying microarray technologies. Nguyen et al. applied high-density cDNA colony filters for a preliminary investigation of differential expression in three cell types present in murine thymus [5] while Schena et al. used microarrays containing 1046 human cDNAs of unknown sequence for the identification of novel heat shock and phorbol ester-regulated genes in human T cells [12]. Heller et al. in 1997 published on microarrays of selected human genes of probable significance in inflammation as well as of genes expressed in peripheral human blood cells. Comparisons between tissue samples of rheumatoid arthritis and inflammatory bowel disease revealed the novel participation of cytokines and chemokines in both diseases [13]. In 1998, Der et al. raised the number of analysed genes to 6800 and determined the mRNA profiles from IFN-α, -β, or -γ treatments of the human fibrosarcoma cell line, HT1080 [14].

These seminal papers boosted both genomic sciences and inflammation research and established a strong interrelationship. Since then, several thousand papers have been published dealing with microarray based analysis of inflammation processes, some of which are detailed in the following chapters of this book. A range of basic to clinical aspects has been addressed, starting with a variety of different materials such as blood, skin, adipose, lung, or brain tissues, as well as sorted or microdissected cells derived from almost all aspects of inflammation processes.

Another discipline which was also subject to intense microarray analysis from the first day onward was cancer research. As early as 1996, DeRisi et al. used a high density microarray of 1161 DNA elements to search for tumorigenic properties of a human melanoma cell line, UACC-903 [15]. Many more life science branches have exploited the advantages of microarrays. They have mainly been applied to gene discovery, transcriptome analysis, functional analysis of new genes, drug validation, and pharmaco- and toxicogenomics [16–25]. The importance of gene discovery lies in the identification of gene expression profiles and correlation with biological states of cells, tissues, or organs during disease or upon drug treatment. This enables the identification of new drug targets but also provides a molecular understanding of disease as reported, for example, by Alizadeh et al.

(2000) where tumor specific expression patterns of hematopoietic (lymphoid) cells where identified [20].

Nowadays, the idea of a miniaturised array of different probes which is reacted with a labelled target to be analysed is used in almost all biological disciplines. Moreover, it is used not only for the analyses of mRNA but for all kinds of classes of molecules such as DNA, proteins, sugar, lipids (see chapter by Gerstmayer for a detailed description).

MicroRNAs (miRNAs) are the newest type of molecule whose expression can be analysed by microarrays. MicroRNAs have been shown to be key regulators of gene expression in many different cellular developmental and physiological processes as divergent as cell lineage decisions, cell proliferation, apoptosis, or morphogenesis and, moreover, inflammation processes. It is challenging to discriminate the short and, in many cases, highly homologous miRNAs by hybridization, especially in a parallel binding reaction on microarrays. To optimize microarray performance, different probe designs and hybridization conditions have been investigated and optimised [26].

As for mRNA analysis, miRNA is also used to understand the basis of inflammation processes. Monticelli et al. used oligonucleotide microarray membranes to analyze the expression of 181 miRNAs in among other, fully differentiated effector cells (Th1 and Th2 lymphocytes and mast cells) and precursors at comparable stages of differentiation (double negative thymocytes and pro-B cells) [27]. In another report, O'Connel et al. used microarray technology to identify miRNAs induced in primary murine macrophages after exposure to polyriboinosinic:polyribocytidylic acid or the cytokine IFN-β [28]. We compared the performance of miRNA microarrays [26] to the sequencing of small RNA libraries, showing a remarkably high correlation between the results generated by the two independent methods [26].

What are microarrays?

Since Southern introduced the blotting technique [29], the hybridisation process has been used in a wide range of techniques for the recognition and quantification of DNAs. A comparable method was soon also used for the analysis of RNAs. A limited number of electrophoretically separated heterogeneous RNA samples were immobilized on a membrane and tested with a single labeled cDNA target – the classic northern blot – focusing on the expression of a distinct gene in different samples. Dot blots were then used to enhance the number of addressable samples. But only the reversal of this procedure – the arraying of multiple homogenous cDNAs and testing with a single heterogeneous labelled sample – made it possible to simultaneously study the gene expression profile of thousand of genes in a given biological sample.

Microarrays are miniaturized devices of a size up to 2.5 × 7.5 cm. They consist of a planar surface where a multitude of different probes have been gritted in a regular fashion at a well defined position. The available microarray formats still differ with respect to DNA probes, supports, and deposition/*in situ* synthesis technologies. DNA molecules are either synthesized *in situ* directly on the surface or pre-synthesized and then deposited. With the *in situ* approach, oligonucleotides are synthesized either by a photolithographic technology or by an ink-jet printer derived deposition of single nucleotides [3, 30, 31]. Nucleotides protected by a chemically- or photo-detachable group are applied one after another to a silicon wafer or glass slide to build a growing oligonucleotide chain. The second technique for producing microarrays is the deposition of prefabricated oligonucleotides or cDNAs pioneered by the Brown lab [6].

Handling of microarrays

Essentially, the handling of microarrays starts with the isolation of RNA from a given tissue or cells. The sample is then labelled either enzymatically by converting the mRNA to cDNAs in an RT-reaction while incorporating labelled nucleotides, or chemically by linking reactive molecules to the RNA. End labelled primers may also be used for cDNA synthesis. The labelled molecules are then hybridised to the microarray. If the incorporated label carries a fluorescent tag, a direct (or one-step) labelling is possible and the hybridised array is ready for read out. In case of an indirect labelling, the incorporated tag has to be contacted by a second (and perhaps third) molecule such as a streptavidin- or antibody-conjugate to produce a signal, and sometimes also signal amplification. The method of labelling determines the sensitivity and hence the minimum amount of starting material for successful hybridisation on the one hand and, on the other, the dynamic range and linearity of the relationship between sample amount and signal intensity. The direct incorporation of fluorescent labelled nucleotides is fast and best understood in terms of robustness and reliability. When using direct fluorescent labelling without any amplification step, the amount of starting material needs to be around 1 to 5 μg of total RNA, which corresponds to 1×10^5–5×10^6 cells or 1–50 mg tissue. Conventional microarray hybridisation processes need around 10^6 fluorescent molecules to yield a detectable signal per spot. As the amount of starting material is most often a limiting factor, various amplification protocols have been developed. Some are performed after hybridisation, amplifying the signal on the array by e.g. enzymatic reactions. Others are based on RNA or cDNA amplifications [32–38].

A critical step in array applications is the hybridisation process. The shorter the probes, the more the hybridisation conditions have to be adjusted and optimised to allow for mismatch detection or to reduce cross-hybridisations. Carefully selected buffer ingredients and a tight control of temperature and pressure must be combined with a low volume and active transport of liquids. These parameters ultimately define the reproducibility, the speed, the sensitivity and selectivity of the assay. In particular, active mass transport is critical, as the hybridisation is often limited by the diffusion rate of the corresponding molecules – most of all in highly viscous solutions. The diffusion rate of e.g. a small DNA molecule is in the range of $< 1\ \mu m^2/s$. That means that it takes several hours for a DNA molecule to move one centimetre before reaching a corresponding spot. As a consequence, complex cDNA probes fail to completely hybridize within a practical time frame. A comparable situation is true for proteins and other biomolecules. To overcome these limitations, several automated hybridisation/incubation stations including active transport of liquid have been developed [38].

Array data: acquisition, analysis and mining

Data acquisition from microarray experiments consists of two steps: the read out of the fluorescence signals by laser scanning devices and the following image analysis using appropriate software packages. Once having generated the primary data, the signal intensities of different samples must be normalized to allow for proper comparison. Several types of array, hybridisation, and dye effects should be considered. This can be accounted for by linear or non-linear normalization methods such as the LOWESS fit originally proposed by Cleveland [39]. The respective iterative function estimates the normalized value for each single spot using the position of the vicinal spots (parameters have to be defined) weighting them according to the distance from the spot of interest.

Reliable identification of candidate genes by statistical methods often suffers from a limited number of replicate experiments. Since users of such approaches are sometimes overwhelmed by the amount of data already produced by performing one single experiment, they ignore the basic necessity of repeat assays. Biological replicates are essential when dealing with expression profiling, especially if subtle changes of gene expression are used to define e.g. disease states or to distinguish substances by means of their impact.

As a result of the multiparametric nature of microarray experiments, bioinformatics and data mining represent essential tools for interpreting the heap of numerical data produced by (series of) microarray experiments. Starting from relatively simple demands for appropriate visualization of the data, bioinformatics tools are necessary to focus on candidate genes and point out subtle changes in expression over many genes. Such expression patterns have predictive power but are difficult to spot. There are two approaches to use microarray data for phenotype prediction: one is to build a predictor based on machine learning techniques and the second is to use microarray analysis to find characteristic marker genes. Suppose you have sets of samples from healthy and diseased tissue that differ slightly in expression and you would like to use these data to predict the medical state of an unknown sample. Following the first approach, tools such as instance-based classifiers or support vector machines compare experimental results against a library of known, classified expression data, or models derived from such data, to predict the medical state of the probe material. The disadvantage is that these predictors act as black boxes, providing an answer but no explanation.

In the second approach, one tries to find interesting groups of genes or single marker genes. The pressing need to make sense of raw microarray data has led to the rapid development of computer-driven analysis tools for this purpose, some of them adapted from the field of data mining, some of them developed specifically for biomedical data of this kind. Several of them have been used in a large number of publications and are well established, most notably cluster analysis [40], self-organizing maps, principal component analysis or SAM (statistical analysis of microarrays) [41].

All of these methods strive to reduce the number of variables by grouping genes based on their expression values. This makes it possible to visualize the results and makes it easier for experts to interpret the data. Also, in analogy to experiments that search for interesting, unknown information, all of these methods are exploratory and unsupervised: they do not make use of existing biological knowledge about the genes they are grouping.

Human experts are still necessary to correctly interpret results, and this interpretation is often the most time consuming and cost intensive step in the whole microarray investigation. A researcher is confronted with a hierarchically clustered tree of several hundred or thousand genes, and he must find the single group that holds the few genes of interest. Usually, this means a lot of time is spent in reading the literature and searching databases.

The logical next step that has taken shape over time is the automation of as much of this expensive work as possible. Ample functional information about genes is available in databases and this knowledge can be used to find groups of genes that are not likely to occur by chance and seem to share biological traits or pathways. Databases e.g. for gene regulatory networks, metabolic networks, protein-protein interaction, and signal transduction networks store information about the interaction between genes. Combining these networks with expression data, it is possible to search for genes that are not only differentially expressed, but also near each other in those biological networks. This is an orthogonal source of evidence that can help us understand what is going on, and statistically support interesting clusters.

Pre-microarray considerations – keep it constant, keep it cold, do it fast

A fundamental assumption for microarray experiments is that every biological process is mirrored by a certain gene expression or change in gene expression. Unfortunately, this is not only true for the experimental state one would like to explore but also for all handling peculiarities prior to microarray hybridisation. Thus, how can we assure that the expression pattern present at the moment we conduct an experiment is still the same when we apply the labelled target to a microarray? While this is a general issue when conducting experiments consisting of multiple steps, it is a rather underestimated one for microarray experiments, especially when inflammation processes are addressed. One reason for this is the fragile nature of RNA. A second one is the complexity of microarray-based studies making it necessary to subdivide the process into several steps which are partly assigned to different labs at different locations.

A distortion of the RNA profile will generally occur as soon as a sample is stressed prior to fixation or labelling. The possibilities of stressing a sample are so manifold that it is hardly possible to give a comprehensive list. There may be "biological stress" for instance when different blood samples are pooled, or shear stress when blood is drawn through a pipette, or when cells or tissue are passed thru a filter or nozzle. But cutting a tissue sample will also generate variable stress on the sample, depending on the material or technique which is used for cutting. These issues are especially critical if microdissection or cell sorting is performed. Further, fixation also has to be monitored cautiously. All kinds of fixation such as cell lysis and RNAse inhibition, or crosslinking using formaldehyde, or freezing will stress the sample differently, depending also on the size and nature of the sample as well as the diffusion rate or progression time of the fixative. The influence of RNases extends to the steps performed after fixation, especially if de-fixation is needed for RNA extraction. This e.g. can be thawing of a sample as well as the dilution of a chemical fixative. Fixation generally leads to at least a partial lysis of cells and, therefore, to a release of intracellular RNases as well as enhanced vulnerability to RNases as soon as a sample is de-fixated. Lastly, different protocols for RNA extraction, e.g. silica or phenol based approaches, but also for mRNA extraction or cDNA synthesis using different oligo(dT) as well as different RNA amplification methods, will result in different gene expression profiles. Therefore, three very simple rules are: keep essentially all parameters, even the weirdest, constant for each set of experiments, keep the samples cool, and perform all sample preparation steps as fast as possible.

Enhanced sensitivity by reducing the complexity of a sample

In the past it has been shown for various chronic diseases that specific gene expression patterns are reflected at the level of whole blood or tissue analysis [42]. This is partly due to a remarkable absolute sensitivity of microarrays which permits starting with a single cell and detecting a few transcripts if appropriate amplification methods are used. However, blood, skin, fat, or brain are complex tissues comprising numerous cell types. Therefore, a substantial proportion of what is usually reported as up-regulation or down-regulation might actually be the result of a shift in cell populations and not in a true regulatory process. Moreover, the contribution of rare cell types to a whole-tissue expression profile might not be detected. A simple calculation helps to illustrate this. Assume one has a sample of 100 cells expressing a given gene at a basal level of 100 transcripts. Let us further assume one of those 100 cells is a T cell being activated by an antigen leading to a 10-fold overexpression of that respective gene. If all 100 cells are analysed, there is only a difference in expression level of 9%, which will not be detect in a commonly performed microarray analysis with, let us say, three replicates. In order to circumvent these problems, several techniques such as flow or magnetic sorting as well as microdissection have been established to analyse purified subpopulations rather than whole-blood or -tissue samples [43, 44] (see respective chapters in this book). Still, as the preparation of PBMCs by, for example, Ficoll gradient is time-consuming, cumbersome, and not amenable to automation, direct whole-blood cell separation and gene expression profiling protocols have been set up [45]. Regarding solid tissues, again, several protocols have been established to standardize enzymatic and mechanical tissue dissociation [46].

Concluding remarks

Microarray technology has evolved into a robust analytical tool which is used routinely in inflammation research. Optimization of their application is currently focusing on the improvement of sample preparation including cell sorting, microdissection, RNA preservation, as well as RNA amplification. In particular, the processing of whole blood and sorted blood cells is still suboptimal. There is, for example, still no available method to preserve the RNA in whole blood without lysing the cells which prohibit their subsequent sorting. Eventually, combining both sample preparation and microarray processing in integrated instruments will terminate microarray technology development.

While the number of reports dealing with microarray application to mRNA profiling has reached a plateau in recent years, other classes of molecules are increasingly being tested on comparable platforms. Most notably, the advent of previously unrecognised RNA classes such as miRNAs offer exiting new possibilities for the understanding of inflammation processes.

References

1 Drmanac R, Labat I, Brukner I, Crkvenjakov R (1989) Sequencing of megabase plus DNA by hybridization: theory of the method. *Genomics* 4, 114–128

2 Khrapko KR, Lysov Yu P, Khorlyn AA, Shick VV, Florentiev VL, Mirzabekov AD

(1989) An oligonucleotide hybridization approach to DNA sequencing. *FEBS Lett* 256, 118–122

3 Fodor SP, Read JL, Pirrung MC, Stryer L, Lu AT, Solas D (1991) Light-directed, spatially addressable parallel chemical synthesis. *Science* 251, 767–773

4 Maskos U, Southern EM (1992) Oligonucleotide hybridizations on glass supports: a novel linker for oligonucleotide synthesis and hybridization properties of oligonucleotides synthesised *in situ*. *Nucleic Acids Res* 20, 1679–1684

5 Nguyen C, Rocha D, Granjeaud S, Baldit M, Bernard K, Naquet P, Jordan BR (1995) Differential gene expression in the murine thymus assayed by quantitative hybridization of arrayed cDNA clones. *Genomics* 29, 207–216

6 Schena M, Shalon D, Davis RW, Brown PO (1995) Quantitative monitoring of gene expression patterns with a complementary DNA microarray [see comments]. *Science* 270, 467–470

7 Bowtell DD (1999) Options available – from start to finish – for obtaining expression data by microarray [published erratum appears in *Nat Genet* 21(2):241]. *Nat Genet* 21, 25–32

8 Tomiuk S, Hofmann K (2001) Microarray probe selection strategies. *Briefings Bioinf* 2, 329–340

9 Lander ES, Linton LM, Birren B, Nusbaum C, Zody MC, Baldwin J, Devon K, Dewar K, Doyle M, FitzHugh W et al (2001) Initial sequencing and analysis of the human genome. *Nature* 409, 860–921

10 Venter JC, Adams MD, Myers EW, Li PW, Mural RJ, Sutton GG, Smith HO, Yandell M, Evans CA, Holt RA et al (2001) The sequence of the human genome. *Science* 291, 1304–1351

11 Brazma A, Hingamp P, Quackenbush J, Sherlock G, Spellman P, Stoeckert C, Aach J, Ansorge W, Ball CA, Causton HC et al (2001) Minimum information about a microarray experiment (MIAME)-toward standards for microarray data. *Nat Genet* 29, 365–371

12 Schena M, Shalon D, Heller R, Chai A, Brown PO, Davis RW (1996) Parallel human genome analysis: microarray-based expression monitoring of 1000 genes. *Proc Natl Acad Sci USA* 93, 10614–10619

13 Heller RA, Schena M, Chai A, Shalon D, Bedilion T, Gilmore J, Woolley DE, Davis RW (1997) Discovery and analysis of inflammatory disease-related genes using cDNA microarrays. *Proc Natl Acad Sci USA* 94, 2150–2155

14 Der SD, Zhou A, Williams BR, Silverman RH (1998) Identification of genes differentially regulated by interferon alpha, beta, or gamma using oligonucleotide arrays. *Proc Natl Acad Sci USA* 95, 15623–15628

15 DeRisi J, Penland L, Brown PO, Bittner ML, Meltzer PS, Ray M, Chen Y, Su YA (1996) Use of a cDNA microarray to analyse gene expression patterns in human cancer. *Nat Genet* 4, 367–70

16 Debouck C, Goodfellow PN (1999) DNA microarrays in drug discovery and development. *Nat Genet* 21, 48–50

17 Farr S, Dunn RT 2nd (1999) Concise review: gene expression applied to toxicology [In Process Citation]. *Toxicol Sci* 50, 1–9

18 Golub TR, Slonim DK, Tamayo P, Huard C, Gaasenbeek M, Mesirov JP, Coller H, Loh ML, Downing JR, Caligiuri MA et al (1999) Molecular classification of cancer: class discovery and class prediction by gene expression monitoring. *Science* 286, 531–537

19 Perou CM, Jeffrey SS, van de Rijn M, Rees CA, Eisen MB, Ross DT, Pergamenschikov A, Williams CF, Zhu SX, Lee JC et al (1999) Distinctive gene expression patterns in human mammary epithelial cells and breast cancers. *Proc Natl Acad Sci USA* 96, 9212–9217

20 Alizadeh AA, Eisen MB, Davis RE, Ma C, Lossos IS, Rosenwald A, Boldrick JC, Sabet H, Tran T, Yu X et al (2000) Distinct types of diffuse large B-cell lymphoma identified by gene expression profiling [see comments]. *Nature* 403, 503–511

21 Bittner M, Meltzer P, Chen Y, Jiang Y, Seftor E, Hendrix M, Radmacher M, Simon R, Yakhini Z, Ben-Dor A et al (2000) Molecular classification of cutaneous malignant melanoma by gene expression profiling. *Nature* 406, 536–540

22 Jain KK (2000) Applications of biochip and microarray systems in pharmacogenomics. *Pharmacogenomics* 1, 289–307

23 Scherf U, Ross DT, Waltham M, Smith LH, Lee JK, Tanabe L, Kohn KW, Reinhold WC, Myers TG, Andrews DT et al (2000) A gene expression database for the molecular pharmacology of cancer [see comments]. *Nat Genet* 24, 236–244

24 Bosio A, Knorr C, Janssen U, Gebel S, Haussmann HJ, Muller T (2002) Kinetics of gene expression profiling in Swiss 3T3 cells exposed to aqueous extracts of cigarette smoke. *Carcinogenesis* 23, 741–748

25 Gebel S, Gerstmayer B, Bosio A, Haussmann HJ, Van Miert E, Muller T (2004) Gene expression profiling in respiratory tissues from rats exposed to mainstream cigarette smoke. *Carcinogenesis* 25, 169–178

26 Landgraf P, Rusu M, Sheridan R, Sewer A, Iovino N, Aravin A, Pfeffer S, Rice A, Kamphorst AO, Landthaler M et al (2007) A mammalian microRNA expression atlas based on small RNA library sequencing. *Cell* 129, 1401–1414

27 Monticelli S, Ansel KM, Xiao C, Socci ND, Krichevsky AM, Thai TH, Rajewsky N, Marks DS, Sander C, Rajewsky K et al (2005) MicroRNA profiling of the murine hematopoietic system. *Genome Biol* 6, R71

28 O'Connell RM, Taganov KD, Boldin MP, Cheng G, Baltimore D (2007) MicroRNA-155 is induced during the macrophage inflammatory response. *Proc Natl Acad Sci USA* 104, 1604–1609

29 Southern EM (1975) Detection of specific sequences among DNA fragments separated by gel electrophoresis. *J Mol Biol* 98, 503–517

30 Fodor SP, Rava RP, Huang XC, Pease AC, Holmes CP, Adams CL (1993) Multiplexed biochemical assays with biological chips. *Nature* 364, 555–556

31 Hughes TR, Mao M, Jones AR, Burchard J, Marton MJ, Shannon KW, Lefkowitz SM, Ziman M, Schelter JM, Meyer MR et al (2001) Expression profiling using microarrays fabricated by an ink-jet oligonucleotide synthesizer. *Nat Biotechnol* 19, 342–347

32 Eberwine J, Yeh H, Miyashiro K, Cao Y, Nair S, Finnell R, Zettel M, Coleman P (1992) Analysis of gene expression in single live neurons. *Proc Natl Acad Sci USA* 89, 3010–3014
33 Ginsberg SD, Che S, Counts SE, Mufson EJ (2006) Single cell gene expression profiling in Alzheimer's disease. *NeuroRx* 3, 302–318
34 Hartmann CH, Klein CA (2006) Gene expression profiling of single cells on large-scale oligonucleotide arrays. *Nucleic Acids Res* 34, e143
35 Iscove NN, Barbara M, Gu M, Gibson M, Modi C, Winegarden N (2002) Representation is faithfully preserved in global cDNA amplified exponentially from sub-picogram quantities of mRNA. *Nat Biotechnol* 20, 940–943
36 Smith L, Underhill P, Pritchard C, Tymowska-Lalanne Z, Abdul-Hussein S, Hilton H, Winchester L, Williams D, Freeman T, Webb S et al (2003) Single primer amplification (SPA) of cDNA for microarray expression analysis. *Nucleic Acids Res* 31, e9
37 Van Gelder RN, von Zastrow ME, Yool A, Dement WC, Barchas JD, Eberwine JH (1990) Amplified RNA synthesized from limited quantities of heterogeneous cDNA. *Proc Natl Acad Sci USA* 87, 1663–1667
38 Appay V, Bosio A, Lokan S, Wiencek Y, Biervert C, Kusters D, Devevre E, Speiser D, Romero P, Rufer N et al (2007) Sensitive gene expression profiling of human T cell subsets reveals parallel post-thymic differentiation for CD4+ and CD8+ lineages. *J Immunol* 179, 7406–7414
39 Yang YH, Dudoit S, Luu P, Lin DM, Peng V, Ngai J, Speed TP (2002) Normalization for cDNA microarray data: a robust composite method addressing single and multiple slide systematic variation. *Nucleic Acids Res* 30, e15
40 Eisen MB, Spellman PT, Brown PO, Botstein D (1998) Cluster analysis and display of genome-wide expression patterns. *Proc Natl Acad Sci USA* 95, 14863–14868
41 Saeed AI, Sharov V, White J, Li J, Liang W, Bhagabati N, Braisted J, Klapa M, Currier T, Thiagarajan M et al (2003) TM4: a free, open-source system for microarray data management and analysis. BioTechniques 34, 374–378
42 Wenzel J, Peters B, Zahn S, Birth M, Hofmann K, Kusters D, Tomiuk S, Baron JM, Merk HF, Mauch C et al (2008) Gene expression profiling of lichen planus reflects CXCL9+-mediated inflammation and distinguishes this disease from atopic dermatitis and psoriasis. *J Invest Dermatol* 128, 67–78
43 Lyons PA, Koukoulaki M, Hatton A, Doggett K, Woffendin HB, Chaudhry AN, Smith KG (2007) Microarray analysis of human leucocyte subsets: the advantages of positive selection and rapid purification. *BMC Genomics* 8, 64
44 Auffray C, Fogg D, Garfa M, Elain G, Join-Lambert O, Kayal S, Sarnacki S, Cumano A, Lauvau G, Geissmann F (2007) Monitoring of blood vessels and tissues by a population of monocytes with patrolling behavior. *Science* 317, 666–670
45 Poggel C, Adams T, Martin S, Pickel C, Prahl N, Schmitz J, Bosio A (2007) Automated sorting of monocytes from whole blood for reproducible gene expression profiling. *American Society of Hematology Annual Meeting Abstracts* 110: 3840
46 Reiß S, Herzig I, Schmitz J, Bosio A, Pennartz S (2008) *Society for Neuroscience Meeting Abstract* 235.7

Methods and protocols for the generation of gene expression profiles in inflammation research

Sample preparation

Whole blood

Birgit Sawitzki and Hans-Dieter Volk

Institute of Medical Immunology, Charité - Universitätsmedizin Berlin, Charitéplatz 1, 10117 Berlin, Germany

Abstract

Within this chapter we summarize the multiple approaches to transcriptional profiling of peripheral blood cells, including the description and characterization of various methodologies. Additionally, we will highlight the advantages and potential pitfalls of these methodologies, especially with regard to application in multi-center clinical trials. Although gene expression profiling after instant whole blood stabilization appears to be superior to expression analysis of isolated cell populations, differences in blood leukocyte composition, high globin RNA content and genomic DNA contamination may affect the results. Thus, the choice of the peripheral blood collection and preparation method must be based on theoretical and practical concerns.

Introduction

Translational medicinal strategies increasingly incorporate gene expression profiling to identify markers or gene expression patterns that are associated with disease pathology or can be used to evaluate a therapeutic benefit or response to a drug [1].

Although transcriptional profiling of the affected tissue would be most desirable, given its relevance to disease mechanisms, in some cases and certain diseases it is impossible to obtain tissue biopsies for gene expression analysis. In these cases, peripheral blood is the only alternative source to conduct gene expression profiling and to identify suitable biomarkers [1]. For animal and clinical studies, minimal invasive procedures would be particularly useful to obtain samples for PCR and DNA-microarray analysis. Peripheral blood is a logical sample material, as it is an accessible biofluid, and circulating leukocytes can contain informative transcripts as a first line of immune defense and sentinels for many disease processes [2]. The continuous interaction between blood cells and the entire body gives rise to the possibility that subtle changes occurring in the context of an injury or a disease, within

Microarrays in Inflammation, edited by Andreas Bosio and Bernhard Gerstmayer

the cells and tissues of the body, may trigger specific changes in gene expression in blood cells reflective of the initiating stimulus [3].

It has been shown that environmental conditions affect transcriptional regulation of tissue-specific genes [3]. Indeed, during the last several years disease specific gene expression patterns have been identified for a variety of human diseases including rheumatoid or psoriatic arthritis [4–7], systemic lupus erythematosus [8], idiopathic thrombocytopenic purpura (ITP) [9], multiple sclerosis [10–15, Crohn's disease [16] and various infectious diseases [17–22] when performing transcriptional profiling on peripheral blood samples. Similar approaches have been performed for many non-inflammatory diseases such as molecular profiling of certain leukemias and lymphomas [23–25]. Changes in peripheral blood gene expression have also been shown to be associated with many other diseases including solid tumors [26, 27], transplant rejection [28–32], hypertension [33, 34], stroke [35–37] and neurological diseases [38]. Additionally, changes in peripheral blood gene expression have been reported upon exposure to biological stimuli such as LPS or tetanus toxoid [39–43], therapeutics including new biologicals [44–46] and environmental exposures [47, 48].

Also, in recent studies it has been successfully shown that peripheral blood transcriptional profiling can be used to predict therapeutic outcome [10, 11, 29, 49]. These results have been obtained in experimental disease models or even as a result of clinical trials [50].

An overview of some recent experimental and clinical reports utilizing peripheral blood gene expression profiling is given in Table 1. This rapidly growing body of evidence demonstrates the potential of using peripheral blood as a surrogate tissue for traditional tissue for diagnosis and prognosis. Clearly, blood provides significant advantages for this purpose, being readily available in large quantities with minimal invasive techniques [3]. Therefore, gene expression changes in leukocytes could possibly serve as early warnings of potential health threats or an early indication of therapy outcome [30]. Yet, to date, few PCR and microarray studies have been performed on peripheral blood sample material, largely because of logistical and technical challenges that accompany collection, storage, processing, and isolation of high quality RNA from this sample source.

Since many clinical trials are multi-center studies, the site-to-site variability of both the collection and processing of blood samples for this purpose must be assessed and taken into consideration before gene expression profiling can be widely adopted.

Here, we will summarize the multiple approaches for transcriptional profiling of peripheral blood cells, including the description and characterization of various methodologies. Additionally, we will highlight the advantages and potential pitfalls of these methodologies, especially with regard to application in multi-center clinical trials. The information provided here is based on a comprehensive review of recent published reports and the authors' own experiences in performing transcriptional profiling of peripheral blood samples in various experimental and clinical settings.

Table 1 - Overview of recent experimental and clinical reports utilizing peripheral blood gene expression profiling

Disease/setting	Sample preparation	Profiling method	Results/summary	Ref.
Exercise	Ery lysis	Microarray, qPCR	450 up- and 150 down-regulated genes following exercise	48
Experimental transplant models	Ery lysis	qPCR	A set of genes whose expression is higher in samples of tolerance developing recipients	29
Rejection of cardiac allografts	Ficoll gradient	Microarray, qPCR	A set of 11 genes which enables discrimination of rejection	31
Liver transplantation	Ficoll gradient	Microarray, qPCR	Gene signature for identification of "tolerant" patients	32
Kidney transplantation	Ficoll gradient, PAXgene	Microarray, qPCR	Gene signature for identification of "tolerant" patients	30
Rheumatoid arthritis subtypes	PAXgene	Microarray	Type I interferon signature in a subpopulation of patients	5
Parkinson	PAXgene	Microarray, qPCR	22 uniquely expressed genes in Parkinson	38
Thrombosis	PAXgene	Microarray	Gene expression pattern characterizing different phenotypes of immune mediated thrombosis	57
Asthma	Ficoll gradient	Microarray, qPCR	Prediction of glucocorticoid responders with a set of 11 to 15 genes	44
LPS challenge	Vacutainer CPT, PAXgene	Microarray, qPCR	Identification of LPS induced and repressed genes	40
Tuberculosis	Roche stabilization reagent	Microarray, qPCR	Set of 9 genes enables identification of patients at risk for recurrent tuberculosis	18
Chronic myeloid leukemia	Ficoll, PAXgene	qPCR	Assessment of minimal residual disease	24

Blood sample preparation methods

A diverse number of platforms for preparing RNA from peripheral blood cells exist, but can be generally classified into two main categories:

1) methodologies that first employ isolation of cell populations or individual leukocyte populations prior to RNA isolation from the purified cells
2) recent methodologies that utilize instant whole blood stabilization and RNA extraction from the whole blood lysate

Both approaches have advantages and potential pitfalls (Tab. 2) and any choice has to be based on the desired analytical performance and logistical possibilities of the clinical study. Before the development of whole blood stabilization reagents, transcriptional profiling was performed on either leukocytes obtained after erythrocyte lysis or PBMC enriched populations.

Ammonium chloride based lysis of erythrocytes is one of the simplest methods to enrich for blood leukocytes. Although easy to perform, the method does not allow standardization, as small changes in blood volume and lysis time can have profound effects on erythrocyte contamination and gene expression patterns. However peripheral gene expression studies have been performed using this sample preparation method in experimental animal models [29, 43].

PBMC separation can be achieved using Ficoll density gradients, Vacutainer™ cell purification tubes or other procedures based on density gradients [21, 31, 40]. These methods do not allow complete isolation of blood leukocytes but rather enable enrichment of monocytes and lymphocytes relative to granulocytes, erythrocytes, reticulocytes and platelets in whole blood. Cell isolation prior to RNA extraction can be extended to purification of individual leukocyte subsets. Indeed,

Table 2 - Advantages and potential pitfalls of different blood preparation methods

	Whole-blood stabilization		Leukocyte isolation
	Paxgene	LeukoLOCK	
RNA stability in unprocessed sample	Yes	Yes	No
Time, equipment and expertise required for processing	Low	Intermediate	High
Sensitivity for transcriptional alterations of low abundant genes	Low	High	High
Ability to select specific cell types	No	No	Yes

magnetic based purification of blood monocytes has been demonstrated to be a useful tool to identify disease specific gene expression profiles [51–53].

Blood samples drawn for erythrocyte lysis, isolation of PBMC or individual leukocyte subsets should be processed immediately at the site of collection or, if not possible, must be stored for a predetermined time prior to processing at a centralized laboratory. As the cellular transcriptome is not stabilized during storage, immediate processing to minimize variability is most desirable [54, 55]. A number of transcripts in peripheral blood are indeed altered within one hour of blood drawing due to contact with foreign material, e.g. plastic [56]. In an attempt to avoid undesired *ex vivo* alterations in the transcriptome, whole blood stabilization reagents have been developed.

Available systems for whole blood RNA stabilization and isolation

Recently, blood RNA isolation reagents have been developed that immediately stabilize RNA upon collection and produce high quality RNA [54]. Their findings highlight that such stabilization reagents prevent *ex vivo* gene induction and RNA degradation due to contact with foreign material or prolonged storage as compared to traditional sampling methods. One such method involves the use of Paxgene blood RNA tubes (PreAnalytiX/Qiagen, Hilden, Germany). The PAXgene system is standardized on BD Vacutainer™ technology that enhances patient and health care worker safety, provides sample protection, and facilitates consistent blood volumes of 2.5 ml per tube on blood collection. This approach avoids the complications inherent in other approaches such as Ficoll density gradient isolation of white cells. Such time-consuming manipulations can significantly alter gene expression patterns compared to data obtained from RNA isolated relatively quickly from freshly collected, stabilized whole blood. Paxgene blood RNA tubes contain a proprietary blend of reagents based on patented RNA stabilization technology [54]. Blend components protect RNA molecules from degradation by RNAses and prevent induction of gene expression for three days at room temperature (18–25°C) and five days at 2–8°C. The stabilized blood may also be frozen in the Paxgene tube at –20°C for prolonged storage or transport. Silica-gel-membrane technology on a column is then used to isolate cellular RNA larger than 50 bp. By performing an on-column DNase I treatment, contamination of the isolated RNA with genomic DNA can be reduced [55]. Successful application of Paxgene tubes for stabilization and gene expression profiling of peripheral blood has been reported for several clinical settings [30, 55, 57, 58].

Another system, Tempus Blood RNA (Applied Biosystems, Foster City, USA), is designed for direct isolation of 3 ml of blood into a plastic, evacuated blood collection tube containing a stabilizing reagent. The RNA is stable in the collection tube for up to five days at room temperature (18–25°C), or longer at 4°C. Similar to the Paxgene tubes, stabilized blood in tempus collection tubes can be frozen at –20°C for long term storage. This system is associated with the ABI 6100 extraction instrument. The stabilizing reagent contains guanidine and detergent. The detergent encapsulates the cellular RNA in a micelle structure and selectively precipitates it, leaving protein and genomic DNA in solution. The lysate is then filtered through a porous membrane that captures RNA larger than 200 bp. During purification, a DNase treatment can be performed on the membrane to remove contaminating genomic DNA. Also, gene expression studies of RNA samples recovered using the Tempus RNA blood system have been successfully performed [23, 59].

RNA/DNA stabilization reagent from Roche Diagnostics (Roche Diagnostics GmbH, Mannheim, Germany) contains guanidinium thiocyanate, Triton-X-100 and a reducing chemical. After addition to the sample, the reagent causes instantaneous lysis of cells and effective inactivation of enzymes such as ribonucleases that would otherwise degrade RNA. EDTA-, citrate- or heparin-treated blood can be used as sample material. Lysates can be stored for twelve months at –15 to –25°C, or for one day at 2–8°C. The stabilization procedure can be combined with an mRNA isolation kit from Roche Diagnostics. The RNA stabilization reagent has been successfully used for the identification of patients at risk for recurrent tuberculosis [18].

RNA isolation of whole blood samples stabilized and preserved by the above mentioned three systems result in a high content of globin mRNA (~70%) from red blood cells (including reticulocytes) interfering with the accurate expression assessment of other genes. Indeed, it has been demonstrated that whole blood RNA samples prepared with these systems result in loss of sensitivity in subsequently performed quantitative RT-PCR and microarray analysis [1, 60]. This situation could compromise detection of gene signatures derived from transcripts of low frequency cell populations such as activated subpopulations of lymphocytes and monocytes [1].

To address this problem, protocols have been established which reduce globin mRNA levels in whole blood cell RNA. Protocols based on two different methodologies have been described. First, an RNase H-dependent digestion of RNA: DNA hemoglobin hybrids has been reported to increase the number of present calls in subsequent microarray analysis [60]. However, due to a lack in specificity the treatment can result in degradation of non-hemoglobin mRNAs and alter the gene expression profile [1]. The second type of strategy employs sequence specific nucleic acid oligomers that are either used to deplete or block hemoglobin mRNA's prior to reverse transcription [61]. Gene expression results obtained after globin mRNA removal will be described in more detail later.

In an attempt to solve the above mentioned problems, Ambion has developed a standardized system (LeukoLOCK) which combines leukocyte enrichment and sample stabilization. This procedure utilizes leukocyte depletion filters that are widely used in blood transfusion therapy. Blood is collected in standard EDTA

Vacutainer™ tubes, and the suction from an empty evacuated blood collection tube is used to draw the sample through the filter in a closed system. The filter captures all leukocyte subsets, including mature myeloid cells, which would be lost during density gradient centrifugation. The filter is then flushed with a solution stabilizing the RNA profile in the intact, captured cells. In contrast, erythrocytes and reticulocytes pass through the filter, resulting in >90% removal of globin mRNA from the isolated RNA. This level of globin mRNA removal is apparently sufficient to rescue low-level gene signals on subsequent analysis without the need for post-extraction globin RNA reduction procedures. The stabilized filter devices can be stored at room temperature (18–25°C) for several days or at –20°C for months.

The system has not been tested for peripheral blood sample preparation and transcriptional profiling in multi-center clinical trials. Since sample preparation and stabilization using the LeukoLOCK system from Ambion requires additional handling steps, its processing may be limited to centralized laboratories.

Influence on RNA quality and quantity

We have highlighted above that careful and thorough design regarding sample collection, stabilization and storage is a prerequisite for disease specific transcriptional profiling of peripheral blood. The amount of specific mRNAs may vary tremendously in both normal and disease settings and influence the accuracy of the measurements. This poses a specific challenge for small, relative changes, particularly with mRNAs of low abundance [62]. Thus, factors such as RNA quality, RNA yield, and the integrity of the reverse transcription reaction are of utmost importance.

Several studies have been undertaken to conduct a systematic comparison between whole blood and leukocyte RNA sampling systems also with regard to RNA yield, stability and quality. Rainen et al. collected parallel blood samples from healthy donors into PAXgene tubes or control EDTA tubes and performed serial RNA extraction on samples stored for five days at room temperature and for up to 90 days at +4 and –20°C [54]. Specific mRNA concentrations in blood stored in EDTA tubes at any temperature decreased over time. Storage in PAXgene tubes prevented this degradation for up to four days at room temperature and at least for 30 days at +4 and –20°C. Storage in EDTA tubes was accompanied by dramatic changes in the expression of individual mRNA species. Expression of, e.g. IL-8 or c-Jun was induced up to 100-fold upon four hours storage at room temperature. These changes in gene expression were prevented when samples were stored in PAXgene tubes.

Kågedal and colleagues have tested PAXgene tubes for detection of minimal residual disease in malignant melanoma [63]. Blood samples of healthy donors were collected into PAXgene tubes, cultured melanoma cells were added and stored at room temperature, +4, and –20°C for one, three, and seven days. Although the overall total RNA integrity versus quality was relatively stable over time, none of their tested transcripts was stable at room temperature. Storage at +4 and -20°C improved stability.

Similar results were obtained by Kim et al. [64]. They evaluated the use of PAX-gene tubes for the detection of gender specific differences in gene expression. RNA integrity was stable for only one day at room temperature, up to four days at +4°C and 194 days at –20°C.

These findings indicate that preservation of whole blood samples using stabilization reagents is limited, and long term storage at –20°C is recommended. The authors' own studies comparing different peripheral blood sample preparation methods have resulted in similar findings. To that end, three different sample collection systems have been compared. 0.5 ml whole blood from healthy subjects was either left unstimulated or *ex vivo* stimulated with 1 µg/ml LPS for four hours. Samples were then divided into three aliquots and prepared using either ammonium chloride lysis, or Paxgene tubes as whole blood collection system and the LeukoLOCK filter. All samples were processed and RNA isolation performed immediately after LPS stimulation. Although all preparation methods resulted in similar RNA quality (see also Tab. 3), RNA yield was lower in samples prepared using the Paxgene whole blood collection system and the LeukoLOCK filter. When reverse transcribed cDNA was evaluated for genomic DNA contamination using an intron specific primer/probe set, Paxgene derived samples contained the highest amount of genomic DNA (C_TINTRON-C_THPRT). These results were obtained despite the fact that the recommended off-column DNase treatment was performed.

Similar findings were reported by Chai et al. [55]. They compared the quality of RNA isolated by erythrocyte lysis or using the PAXgene collection system. Despite on-column DNase I digestion, RNA samples isolated using the PAXgene collection system contained residual amounts of genomic DNA. A second off-column DNase I treatment was sufficient to remove any residual genomic DNA. Such a second DNase treatment may be essential in gene expression studies in which the target gene is expressed at low levels.

Table 3 - Impact of different whole blood preparation methods on RNA quality and quantity

	RNA quantity (µg/ ml blood)	RNA quality (RIN)	Intron specific C_T	HPRT C_T	Foxp3 C_T
Paxgene	1.53 ± 0.54	7.8±2.3	35.4±1.6	31.7±1.1	34.8±1.6
Ery lysis	2.87±0.53	7.5±2.8	34.7±1.3	26.5±0.6	29.7±1.2
LeucoLOCK	1.78±0.66	8.3±1.8	36.8±1.7	27.9±0.9	31.1±1.3

Gene expression profiling results comparing PBMC and stabilized whole blood samples

Feezor et al. evaluated the reliability of PBMC and whole blood preparation methods (PAXgene) [65]. They obtained samples from healthy subjects and trauma patients and subjected them to different RNA extraction protocols using either the traditional leukocyte (buffy coat) isolation or a whole blood collection system. To simulate changes in gene expression that would be anticipated following microbial infection, blood samples from healthy subjects were stimulated *ex vivo* with SEB. Affymetrix U95A and U133A microarrays hybridized with cRNA generated from whole blood collection had consistently fewer present calls as compared to arrays hybridized with cRNA samples of leukocyte preparations. However, the reproducibility was extremely high in samples obtained from both preparation systems. Their analysis also revealed that the RNA isolation method resulted in greater differences in gene expression than stimulation with SEB. The signal-to-noise ratio of the difference between SEB stimulated and unstimulated samples was significantly higher in leukocyte derived samples compared to whole blood preparations. At a high level of significance, twice as many probe sets discriminated between SEB stimulated and unstimulated samples when using leukocyte samples compared to Paxgene samples.

Talwar and colleagues have studied gene expression profiles of peripheral blood leukocytes after endotoxin challenge in humans using leukocyte enriched as well as whole blood samples [40]. They were able to identify an LPS-induced signature which was similar between both preparation methods. However, a reduced number of induced and repressed genes were identified using whole blood prepared samples.

We have performed quantitative RT-PCR based gene expression profiling of samples obtained from more than ten healthy individuals prepared using a Ficoll based leukocyte isolation (1 × 5 ml) and whole blood stabilization tubes (PAXgene, 2 × 2.5 ml) with a particular focus on quantifying the expression of relatively low abundant leukocyte-derived genes such as Foxp3 and Rhamm. Furthermore, although similar amounts of total RNA were used for reverse transcription, PAXgene derived samples had the highest C_T values compared to the leukocyte derived samples (Tab. 4). Additionally, very low abundant genes, such as Rhamm, were not detectable in whole blood stabilized samples.

Table 4 - PCR results comparing Ficoll enriched PBMCs with different whole blood preparation methods

	HPRT, C_T	Foxp3, C_T	Rhamm, C_T
Paxgene	27.8 ± 1.5	30.9 ± 1.2	35.5 ± 1.7
Ficoll	24.5 ± 1.1	27.6 ± 0.9	33.8 ± 1.5

These findings highlight the fact that cell preparation and RNA isolation from peripheral blood is a critical variable when designing clinical trials utilizing whole genome transcription analysis.

Comparability of different whole blood isolation systems

In a study performed by Prezeau, two whole blood stabilizing reagents (PAXgene and Tempus) were compared with regard to their potential use in the assessment of minimal residual disease of leukemia [23]. Both stabilization reagents were also compared to a Ficoll based sample preparation. They reported a lower RNA yield when using the Tempus blood system. They further showed that samples prepared with both whole blood stabilizing methods amplify with the same sensitivity, which was slightly lower than that of sample from purified leukocytes. However, as the ratio of fusion and control gene transcripts was not altered and the detection of minimal residual genes was more stable in whole blood samples over time, both systems could be used for minimal residual follow-up in myeloproliferative and acute leukemias.

When we compared samples prepared using erythrocyte lysis, PAXgene and the LeukoLOC filter with regard to quantifying low, abundantly expressed genes using quantitative RT-PCR, both erythrocyte lysis and the LeukoLOC filter were superior to the PAXgene system (Table 3). Although the expression levels for genes such as Foxp3 were not altered, PAXgene derived samples amplified with a lower sensitivity. But as erythrocyte lysis and the LeukoLOCK filter require additional handling time, equipment, and expertise upon sample collection, the use of PAXgene tubes may be more appropriate when performing multi-center clinical trials. For special applications such as detection of minimal residual disease genes or low abundant genes, more sensitive sample collection and preparation methods should be used.

Hemoglobin RNA reduction

Several groups have studied the effect of globin mRNA reduction of peripheral blood RNA samples on sensitivity and performance of microarray analysis [59–62, 65–68]. Debey and colleagues employed the globin mRNA reduction protocol recommended by Affymetrix. RNA samples are incubated with globin-specific oligos. Globin mRNA-DNA hybrids are removed by adding RNAse H. Non-depleted and depleted samples were tested on GeneChip® Human Genome U133A 2.0 arrays (Affymetrix). Globin mRNA reduction resulted in improved microarray results such as increased present calls and reduced variance. They also demonstrated that this method increases sensitivity to overall small differences in gene expression profiles such as observed between females and males. Four-hundred-and-fifteen transcripts displayed higher mean signal intensity in non-depleted samples as compared to depleted samples while, for nine genes, these differences were statistically significant [60].

In the study by Feezor and colleagues, globin mRNA depletion using RNAse H based degradation was also associated with an increased number of present calls. However, removing globin mRNA species did not result in an identical SEB induced response pattern as compared to that of samples prepared from leukocytes [65].

Field et al. have applied the GLOBINclear™ – human whole blood globin reduction kit from Ambion – to reduce the content of globin mRNAs in samples prepared using the Paxgene blood RNA system. The kit employs nucleic acid hybridization combined with biotin/streptavidin binding and magnetic separation. They report an average recovery of 76% after globin mRNA depletion. Non-depleted and depleted samples were also tested on GeneChip® Human Genome U133A 2.0 arrays (Affymetrix). Globin mRNA depletion resulted in an increase in present calls from an average of 9712 ± 525 genes in non-depleted samples to $12\,365 \pm 258$ genes in depleted samples. All unmasked genes fell into the lowest 10% of the overall range in expression. However, five genes were masked following globin mRNA depletion as they were detectable in non-depleted samples but almost absent in depleted samples [61].

To assess the influence of globin mRNA on genome wide expression studies, we studied the gene expression profile of 14 RNA samples, isolated using the Paxgene system, prior to and after hemoglobin mRNA depletion. Hemoglobin mRNA depletion was performed using the GLOBINclear™ kit. RNA recovery ranged from 46% and 83% of the processed total RNA amount. Non-depleted and depleted samples were amplified and the quality of the resulting amplified aRNA samples was analyzed on an Agilent Bioanalyzer. Depletion of hemoglobin mRNA led to a significant improvement in size and distribution of the amplified RNA (Fig. 1). Amplified RNA samples were converted into Cy5- and Cy3-fluorescently labeled cDNA and hybridized onto custom designed PIQOR™ microarrays (Miltenyi Biotec). Comparison of the gene expression profiles before and after depletion revealed a 1.7-fold increase in the number of detectable genes, confirming the negative influence of hemoglobin mRNA on the sensitivity of microarray experiments. Global correlation analysis revealed moderate to good correlation values between non-depleted and depleted samples. A discriminatory gene analysis revealed 12 discriminatory genes for depleted *versus* non-depleted samples. Most of the discriminatory genes were upregulated in non-depleted samples, which points to an additional systematic influence of hemoglobin mRNA depletion on gene expression profiling. A possible explanation might be that the samples were not only depleted of globin but also of additional genes (unpublished observations).

In summary, we can conclude, as the increase in present calls outweighs the masking of a few gene transcripts, globin mRNA reduction seems to be beneficial for gene expression profiling.

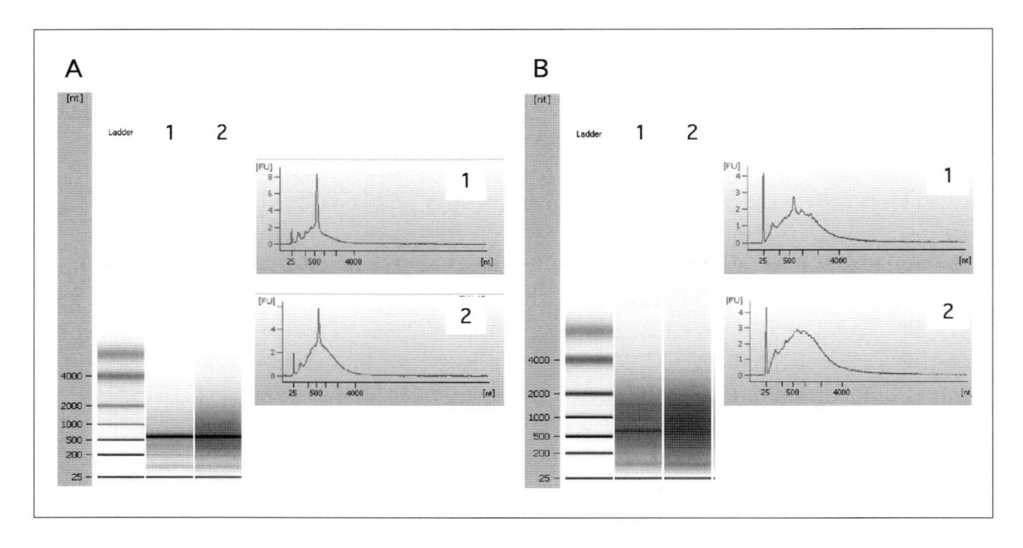

Figure 1.
Effect of hemoglobin mRNA removal of microarray analysis.
Hemoglobin mRNA depletion was performed using the GLOBINclear™ kit. Non-depleted and depleted samples were amplified and the quality of the resulting amplified aRNA samples was analyzed on an Agilent Bioanalyzer.

Concluding remarks

Whole blood collection systems such as PAXgene offer a number of technical advantages that make them highly attractive for multicenter clinical trials. As gene expression profiling using whole blood stabilized samples results in decreased sensitivity and signal-to-noise ratio, study design will require replicate sampling and a higher number of study patients to identify genes whose expression may discriminate diseases, treatment responses, and outcome. Thus, the choice of the peripheral blood collection and preparation method must be based on theoretical and practical concerns.

References

1 Burczynski, M.E. et al. Transcriptional profiling of peripheral blood cells in clinical pharmacogenomic studies. *Pharmacogenomics* 7, 187–202 (2006)
2 Whitney, A.R. et al. Individuality and variation in gene expression patterns in human blood. *Proc Natl Acad Sci USA* 100, 1896–901 (2003)

3 Liew, C.C. et al. The peripheral blood transcriptome dynamically reflects system wide biology: a potential diagnostic tool. *J Lab Clin Med* 147, 126–32 (2006)
4 Batliwalla, F.M. et al. Microarray analyses of peripheral blood cells identifies unique gene expression signature in psoriatic arthritis. *Mol Med* 11, 21–9 (2005)
5 van der Pouw Kraan, T.C. et al. Expression of a pathogen-response program in peripheral blood cells defines a subgroup of Rheumatoid Arthritis patients. *Genes Immun* 9, 16–22 (2007)
6 Edwards, C.J. et al. Molecular profile of peripheral blood mononuclear cells from patients with rheumatoid arthritis. *Mol Med* 13, 40–58 (2007)
7 Ogilvie, E.M. et al. Specific gene expression profiles in systemic juvenile idiopathic arthritis. *Arthritis Rheum* 56, 1954–65 (2007)
8 Alcorta, D.A. et al. Leukocyte gene expression signatures in antineutrophil cytoplasmic autoantibody and lupus glomerulonephritis. *Kidney Int* 72, 853–64 (2007)
9 Sood, R. et al. Gene expression profile of idiopathic thrombocytopenic purpura (ITP) *Pediatr Blood Cancer* 47, 675–7 (2006)
10 Gilli, F. et al. Biological markers of interferon-beta therapy: comparison among interferon-stimulated genes MxA, TRAIL and XAF-1. *Mult Scler* 12, 47–57 (2006)
11 Gilli, F. et al. Qualitative and quantitative analysis of antibody response against IFNbeta in patients with multiple sclerosis. *Mult Scler* 12, 738–46 (2006)
12 Achiron, A. et al. Peripheral blood gene expression signature mirrors central nervous system disease: the model of multiple sclerosis. *Autoimmun Rev* 5, 517–22 (2006)
13 Achiron, A. et al. Impaired expression of peripheral blood apoptotic-related gene transcripts in acute multiple sclerosis relapse. *Ann NY Acad Sci* 1107, 155–67 (2007)
14 Fossey, S.C. et al. Identification of molecular biomarkers for multiple sclerosis. *J Mol Diagn* 9, 197–204 (2007)
15 Singh, M.K. et al. Gene expression changes in peripheral blood mononuclear cells from multiple sclerosis patients undergoing beta-interferon therapy. *J Neurol Sci* 258, 52–9 (2007)
16 Burczynski, M.E. et al. Molecular classification of Crohn's disease and ulcerative colitis patients using transcriptional profiles in peripheral blood mononuclear cells. *J Mol Diagn* 8, 51–61 (2006)
17 Fjaerli, H.O. et al. Whole blood gene expression in infants with respiratory syncytial virus bronchiolitis. *BMC Infect Dis* 6, 175 (2006)
18 Mistry, R. et al. Gene-expression patterns in whole blood identify subjects at risk for recurrent tuberculosis. *J Infect Dis* 195, 357–65 (2007)
19 Wiersinga, W.J. et al. High-throughput mRNA profiling characterizes the expression of inflammatory molecules in sepsis caused by *Burkholderia pseudomallei*. *Infect Immun* 75, 3074–9 (2007)
20 Jacobsen, M. et al. Candidate biomarkers for discrimination between infection and disease caused by Mycobacterium tuberculosis. *J Mol Med* 85, 613–21 (2007)
21 Lempicki, R.A. et al. Gene expression profiles in hepatitis C virus (HCV) and HIV

coinfection: class prediction analyses before treatment predict the outcome of anti-HCV therapy among HIV-coinfected persons. *J Infect Dis* 193, 1172–7 (2006)

22 Kawada, J. et al. Analysis of gene-expression profiles by oligonucleotide microarray in children with influenza. *J Gen Virol* 87, 1677–83 (2006)

23 Prezeau, N. et al. Assessment of a new RNA stabilizing reagent (Tempus Blood RNA) for minimal residual disease in onco-hematology using the EAC protocol. *Leuk Res* 30, 569–74 (2006)

24 Thorn, I. et al. The impact of RNA stabilization on minimal residual disease assessment in chronic myeloid leukemia. *Haematologica* 90, 1471–6 (2005)

25 Sakhinia, E. et al. Comparison of gene-expression profiles in parallel bone marrow and peripheral blood samples in acute myeloid leukaemia by real-time polymerase chain reaction. *J Clin Pathol* 59, 1059–65 (2006)

26 Twine, N.C. et al. Disease-associated expression profiles in peripheral blood mononuclear cells from patients with advanced renal cell carcinoma. *Cancer Res* 63, 6069–75 (2003)

27 Burczynski, M.E. et al. Transcriptional profiles in peripheral blood mononuclear cells prognostic of clinical outcomes in patients with advanced renal cell carcinoma. *Clin Cancer Res* 11, 1181–9 (2005)

28 Alakulppi, N.S. et al. Diagnosis of acute renal allograft rejection by analyzing whole blood mRNA expression of lymphocyte marker molecules. *Transplantation* 83, 791–8 (2007)

29 Sawitzki, B. et al. Identification of gene markers for the prediction of allograft rejection or permanent acceptance. *Am J Transplant* 7, 1091–102 (2007)

30 Brouard, S. et al. Identification of a peripheral blood transcriptional biomarker panel associated with operational renal allograft tolerance. *Proc Natl Acad Sci USA* 104, 15448–53 (2007)

31 Deng, M.C. et al. Noninvasive discrimination of rejection in cardiac allograft recipients using gene expression profiling. *Am J Transplant* 6, 150–60 (2006)

32 Martinez-Llordella, M. et al. Multiparameter immune profiling of operational tolerance in liver transplantation. *Am J Transplant* 7, 309–19 (2007)

33 Marteau, J.B. et al. Genetic determinants of blood pressure regulation. *J Hypertens* 23, 2127–43 (2005)

34 Timofeeva, A.V. et al. Altered gene expression pattern in peripheral blood leukocytes from patients with arterial hypertension. *Ann NY Acad Sci* 1091, 319–35 (2006)

35 Baird, A.E. Blood genomic profiling: novel diagnostic and therapeutic strategies for stroke? *Biochem Soc Trans* 34, 1313–7 (2006)

36 Baird, A.E. Blood genomics in human stroke. *Stroke* 38, 694–8 (2007)

37 Moore, D.F. et al. Using peripheral blood mononuclear cells to determine a gene expression profile of acute ischemic stroke: a pilot investigation. *Circulation* 111, 212–21 (2005)

38 Scherzer, C.R. et al. Molecular markers of early Parkinson's disease based on gene expression in blood. *Proc Natl Acad Sci USA* 104, 955–60 (2007)

39 Stordeur, P. et al. Immune monitoring in whole blood using real-time PCR. *J Immunol Methods* 276, 69–77 (2003)

40 Talwar, S. et al. Gene expression profiles of peripheral blood leukocytes after endotoxin challenge in humans. *Physiol Genomics* 25, 203–15 (2006)

41 Fannin, R.D. et al. Differential gene expression profiling in whole blood during acute systemic inflammation in lipopolysaccharide-treated rats. *Physiol Genomics* 21, 92–104 (2005)

42 Calvano, S.E. et al. A network-based analysis of systemic inflammation in humans. *Nature* 437, 1032–7 (2005)

43 Brownstein, B.H. et al. Commonality and differences in leukocyte gene expression patterns among three models of inflammation and injury. *Physiol Genomics* 24, 298–309 (2006)

44 Hakonarson, H. et al. Profiling of genes expressed in peripheral blood mononuclear cells predicts glucocorticoid sensitivity in asthma patients. *Proc Natl Acad Sci USA* 102, 14789–94 (2005)

45 Potti, A. et al. Genomic signatures to guide the use of chemotherapeutics. *Nat Med* 12, 1294–300 (2006)

46 Lequerre, T. et al. Gene profiling in white blood cells predicts infliximab responsiveness in rheumatoid arthritis. *Arthritis Res Ther* 8, R105 (2006)

47 Rokutan, K. et al. Gene expression profiling in peripheral blood leukocytes as a new approach for assessment of human stress response. *J Med Invest* 52, 137–44 (2005)

48 Buttner, P. et al. Exercise affects the gene expression profiles of human white blood cells. *J Appl Physiol* 102, 26–36 (2007)

49 Haider, A.S. et al. Novel insight into the agonistic mechanism of alefacept *in vivo*: differentially expressed genes may serve as biomarkers of response in psoriasis patients. *J Immunol* 178, 7442–9 (2007)

50 Hagberg, A. et al. Gene expression analysis identifies a genetic signature potentially associated with response to alpha-IFN in chronic phase C ml patients. *Leuk Res* 31, 931–8 (2007)

51 Lyons, P.A. et al. Microarray analysis of human leukocyte subsets: the advantages of positive selection and rapid purification. *BMC Genomics* 8, 64 (2007)

52 Tang, Y. et al. Gene expression in blood changes rapidly in neutrophils and monocytes after ischemic stroke in humans: a microarray study. *J Cereb Blood Flow Metab* 26, 1089–102 (2006)

53 Du, X. et al. Genomic profiles for human peripheral blood T cells, B cells, natural killer cells, monocytes, and polymorphonuclear cells: comparisons to ischemic stroke, migraine, and Tourette syndrome. *Genomics* 87, 693–703 (2006)

54 Rainen, L. et al. Stabilization of mRNA expression in whole blood samples. *Clin Chem* 48, 1883–90 (2002)

55 Chai, V. et al. Optimization of the PAXgene blood RNA extraction system for gene expression analysis of clinical samples. *J Clin Lab Anal* 19, 182–8 (2005)

56 Hartel, C. et al. *Ex vivo* induction of cytokine mRNA expression in human blood samples. *J Immunol Methods* 249, 63–71 (2001)
57 Potti, A. et al. Gene-expression patterns predict phenotypes of immune-mediated thrombosis. *Blood* 107, 1391–6 (2006)
58 Wang, Y. et al. Gene expression signature in peripheral blood detects thoracic aortic aneurysm. *PLoS ONE* 2, e1050 (2007)
59 Shou, J. et al. Optimized blood cell profiling method for genomic biomarker discovery using high-density microarray. *Biomarkers* 10, 310–20 (2005)
60 Debey, S. et al. A highly standardized, robust, and cost-effective method for genome-wide transcriptome analysis of peripheral blood applicable to large-scale clinical trials. *Genomics* 87, 653–64 (2006)
61 Field, L.A. et al. Functional identity of genes detectable in expression profiling assays following globin mRNA reduction of peripheral blood samples. *Clin Biochem* 40, 499–502 (2007)
62 Ovstebo, R. et al. Quantification of relative changes in specific mRNAs from frozen whole blood - methodological considerations and clinical implications. *Clin Chem Lab Med* 45, 171–6 (2007)
63 Kagedal, B. et al. Failure of the PAXgene Blood RNA System to maintain mRNA stability in whole blood. *Clin Chem Lab Med* 43, 1190–2 (2005)
64 Kim, S.J. et al. Effects of storage, RNA extraction, genechip type, and donor sex on gene expression profiling of human whole blood. *Clin Chem* 53, 1038–45 (2007)
65 Feezor, R.J. et al. Whole blood and leukocyte RNA isolation for gene expression analyses. *Physiol Genomics* 19, 247–54 (2004)
66 Liu, J. et al. Effects of globin mRNA reduction methods on gene expression profiles from whole blood. *J Mol Diagn* 8, 551–8 (2006)
67 McPhail, S. et al. Overcoming challenges of using blood samples with gene expression microarrays to advance patient stratification in clinical trials. *Drug Discov Today* 10, 1485–7 (2005)
68 Fan, H. et al. The transcriptome in blood: challenges and solutions for robust expression profiling. *Curr Mol Med* 5, 3–10 (2005)

Sample preparation

Separation of whole blood cells and its impact on gene expression

Andreas Grützkau and Andreas Radbruch

German Arthritis Research Center (DRFZ), Charitéplatz 1, 10117 Berlin, Germany

Abstract

Ease of sample collection predestines peripheral blood cells, and their transcriptional or translational products, to become surrogate markers for inflammatory processes in several diseases, including cancer, autoimmune, genetic or metabolic disorders. Therefore, peripheral blood mononuclear cells (PBMCs) and whole blood have been commonly used for genome-wide expression analyses. In comparison to whole blood, which primarily consists of erythrocytes, reticulocytes, platelets, granulocytes, T and B lymphocytes, NK cells and monocytes, PBMCs were pre-enriched for lymphocyte populations, NK cells, and monocytes by density gradient centrifugation, such as Ficoll or Percoll. But the cellular composition of blood shows inter-individual variations and is intensely influenced by pathophysiological processes, such as inflammation. Thus, success in terms of reproducibility and interpretability of a microarray experiment greatly depends on samples being comparable in quality and in quantitative cellular composition.

In this context, some theoretical considerations will be bestowed upon problems arising from samples of heterogeneous composition. To overcome these limitations, cell sorting strategies will be presented that have been optimized with respect to the special requirements necessary for global gene expression studies.

Introduction

"To sort, or not to sort" – that is the question to be asked before starting a microarray experiment.

The initial euphoria over estimating the expression of almost all genes simultaneously in a single experiment was soon dampened by the impression to produce irreproducible data sets [1, 2]. Regarding this disappointment, several guidelines were published which focus on a standardisation of experimental parameters that determine the success or failure of a microarray experiment [3–6]. The heterogeneity of a sample with regard to its cellular composition is one of the most underestimated sources of trouble when trying to compare and interpret expression data generated from the same disease in different laboratories using different biological materials.

Microarrays in Inflammation, edited by Andreas Bosio and Bernhard Gerstmayer

Therefore, to plan a microarray experiment the following aspects should be taken into account with respect to estimating the heterogeneity of a biological sample and deciding whether enrichment or purification of an appropriate leukocyte population will or will not be necessary:

- What is the main focus of the study?
- How many replicates will be necessary?
- How much blood or tissue and how many cells will be available and how much RNA can be expected?
- Pooling of samples or amplification of RNA if sample material is limited?

The *main focus* of a microarray experiment will determine whether whole blood, PBMCs or purified cell populations are applicable. In principal, variations in cellular composition will influence the extent of biological variability in a particular sample. An overview of sources of variability with respect to the interpretation of gene expression profiling data has been summarized in several reviews [7, 8]. Consequently, the magnitude of heterogeneity in a sample will affect the *number of replicates* necessary for an array experiment. When comparing *cross-sectionally* and *longitudinally* designed clinical studies it becomes clear that the latter need lower sample numbers because the same subject will be followed at different time points in a clinical trial and possible cellular variations may be mainly induced by the treatment itself. The applicability of whole blood may be sufficient for clinical studies aimed primarily at identifying disease classifying gene patterns, as long as variations in gene expression patterns observed in the blood of healthy subjects is strikingly smaller than the variation ascertained among samples from patients such as in cancer or infection [9]. The influence of inter-individual variations caused by differences in the cellular composition was extensively analysed by global gene expression analyses in 75 healthy volunteers [9]. In this study a significant inter-individual cellular variability was observed which, in part, was responsible for the overall variations seen in individual gene expression patterns. Thus, inter-individual variability may be more relevant than technically-induced differences possibly introduced by sample processing procedures [10].

But analysing whole blood or PBMC samples will be practically unacceptable for purposes primarily intended to explore disease-associated pathomechanisms, since data sets dominated by a high degree of variability may be responsible for detecting only the tip of the iceberg of the total number of differentially expressed genes. In this situation, where special emphasis is placed on a functional interpretation of gene expression patterns, the analysis of purified cell populations is strongly recommended. This allows for clearly assessing whether an apparently changed expression rate of a particular gene is either the result of a transcriptional up- or downregulation or is caused by a change in cellular composition by migratory processes, proliferation or apoptosis/necrosis of cells. Furthermore, it should be noted that major

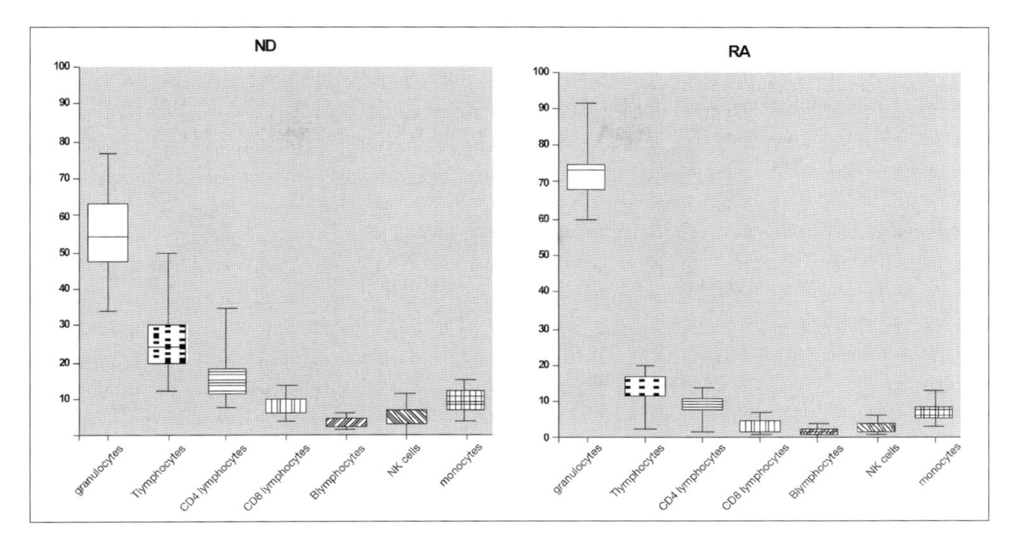

Figure 1.
These boxplots show the cellular composition of whole blood for a group of 20 healthy donors in comparison to a group of 20 patients suffering from active rheumatoid arthritis (RA) which represents a typical chronic-inflammatory disease. The inter-individual variations are given with respect to a particular blood cell population. Minor populations such as NK-cells or B lymphocytes may vary in the range of up to ten-fold. Under inflammatory conditions, in the group of RA patients a significant increase in the proportion of granulocytes is obvious with a mean increase of 35% if compared to healthy donors and may reach highest proportions of up to 90% of total leukocytes.

changes in gene expression levels of a rare cell population might have been lost entirely because they were overshadowed by the background noise of the majority of unchanged cells. This scenario may come true in acute inflammation, where the proportion of granulocytes may increase up to 90% of total leukocytes and cells such as B lymphocytes, NK cells, regulator T lymphocytes or plasma cells are completely underrepresented. In Figure 1 the cellular composition of whole blood is shown for a group of 20 healthy donors in comparison to a group of 20 patients suffering from active rheumatoid arthritis (RA), which represents a typical chronic-inflammatory disease [11]. These graphs give an impression of the inter-individual variation with respect to single blood cell populations and may vary in the range of up to ten-fold, especially with respect to minor populations, such as NK cells. Under inflammatory conditions, in the group of RA patients, a significant increase in the proportion of granulocytes is obvious, with a mean increase of 35% when compared to healthy donors, and may reach the highest proportions of up to 90% of total leukocytes.

Finally, for a functional understanding, it will be important to assign a particular differentially expressed gene to the corresponding cell type, something that is presently an insurmountable barrier with regard to complex cell mixtures. As outlined by *The Tumor Analysis Best Practices Working Group* [12], special, trained computer software will be necessary to recognise the expression profiles for each individual cell type within a mixed sample, so that it should be possible to dissect a complex expression profile into its single, cell-specific constituents.

The decision to sort a cell population of interest for the analysis of its gene expression profile also depends on the *number of cells* that will be available after sorting. Either the amount of blood or the rare cell populations within the scope of interest are the limiting factors. In both cases it might be difficult to enrich cells such that enough RNA material can be obtained, which is necessary for their processing according to standard procedures (e.g. cRNA by one round of *in vitro* transcription). In this case samples from different patients or experiments can be pooled or mRNA amplification strategies must be applied, as discussed elsewhere in this book by S. Ginsberg and C. Klein.

Pooling of samples must be avoided if an estimation of variance within a set of samples is of primary interest [13, 14] but, on the other hand, *amplification procedures*, such as the Small Sample Protocol from Affymetrix, will induce an undefined bias leading to an over- or underestimation of transcripts which, in the end, will not allow comparisons between standard-processed and additionally amplified samples [15].

In view of all points discussed so far, it can be concluded that cell sorting is the silver-bullet solution for genuine and more sensitive gene expression analysis.

The best cell sorting "cocktail": on ice, but without Ficoll

Changes detected in the transcriptome of a biological sample under certain conditions are the sum of real biology-related events and sample-handling induced factors. Since mRNA is, in general, a very instable intermediate and its transcription is influenced by stress signals, such as alterations of body-temperature and oxygen supply, within minutes, a fast sample processing time must be the goal. In this context, it has been demonstrated that even short delays between blood being taken and the start of the separation process leads to significant systematic variations in gene expression in cells of myeloid lineage [9, 10, 16]. This must be taken into account if shipping of samples is necessary.

For this very reason, cell sorting procedures are often rejected for reason of the pitfalls discussed above, and it will be a great challenge to adapt standard cell sorting procedures to the requirements of global gene expression profiling. Up to now, there have only been a few studies available that address this problem in detail [10, 16–18].

In Figure 2 a specially designed experimental setup is shown that we have used to explore the influence of different cell sorting procedures on the transcriptome of peripheral blood monocytes and CD4 T helper lymphocytes. Monocytes are known

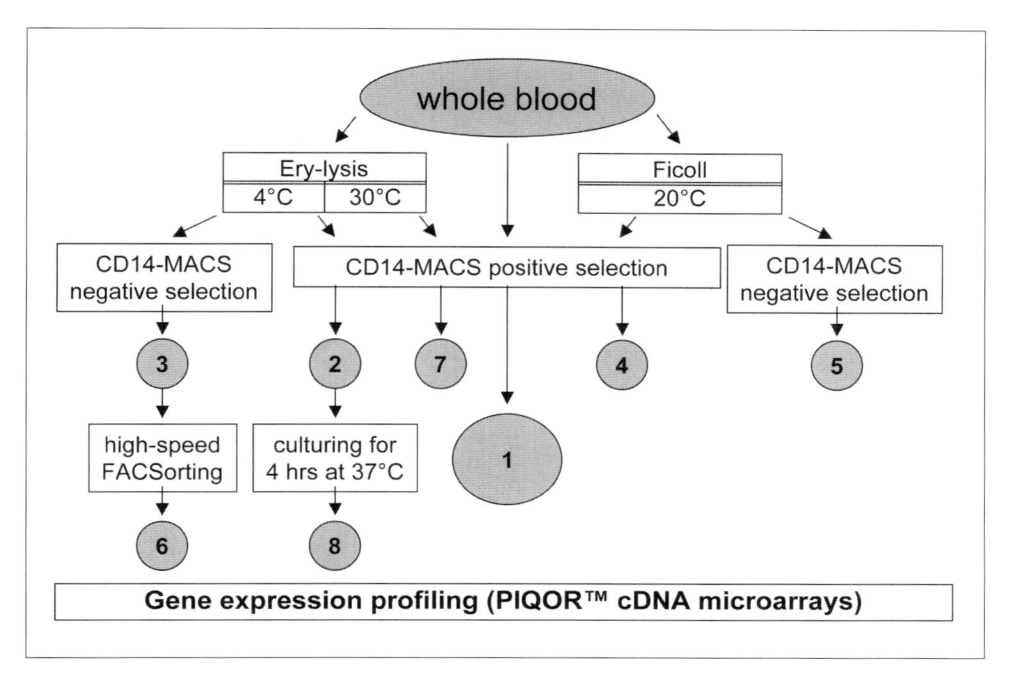

Figure 2.
In figure 2 the experimental setup is schematically shown to explore the influence of different cell sorting procedures on the transcriptome of peripheral blood monocytes and CD4 T helper lymphocytes. Cells obtained by different cell sorting procedures were numbered as indicated. The transcriptional responses of the cell preparations were comparatively analyzed by screening 900 immunologically relevant genes with a custom-made cDNA microarray (PIQOR^{TM} cDNA arrays, Miltenyi Biotec GmbH). Further details are indicated in the text.

to respond very sensitively to *ex vivo* manipulations (such as adherence) which induce a set of early response genes. The following problems concerning purity of cells and the artificial transcriptional alterations possibly induced thereby have been addressed by this approach:

- Comparison of positive and negative (untouched) MACS-sorting techniques
- Comparison of erythrocyte lysis at 4°C and 37°C
- Influence of Ficoll density gradient centrifugation
- Influence of high-speed FACS-sorting

Fluorescence activated and magnetic cell sorting (FACS/MACS) are the most commonly used cell sorting techniques applied to the separation of blood samples and will be discussed in more detail elsewhere in this book.

Hypo-osmotic lysis of erythrocytes (populations # 2, 3 and 7) and Ficoll density gradient centrifugation (populations # 4 and 5) are usually applied to deprive blood samples of erythrocytes alone or along with granulocytes. Both methods were routinely performed at room temperature to ensure effective lysis of erythrocytes and to avoid clumping of cells during density centrifugation. Here, we have used a commercially available erythrocyte lysis buffer that can be used alternatively at 4°C (EL-buffer; QIAGEN GmbH) without any limitations regarding its potency to lyse erythrocytes quantitatively. In order to test the influence of high-speed FACS-sorting we have used negatively enriched cells by using magnetic microbeads (population # 3) and applied half of them to a high-speed FACS-sorting procedure (population # 6).

All eight cell populations obtained were finally analysed by screening 900 immunologically relevant genes with a custom-made cDNA microarray (PIQOR™ cDNA microarrays, Miltenyi Biotec GmbH). Monocytes obtained directly from whole blood by positive selection *via* CD14-labeled magnetic microbeads (population # 1) were used as reference sample, since this is the fastest and most straight-forward way possible to obtain purified monocytes within 20 min. after blood puncture. Additionally, the gene expression profile of monocytes activated *in vitro* by culturing them for 4 h at 37°C was used to achieve a maximal unspecific stimulation (population # 8).

The major conclusion from this study was that lysis of erythrocytes at 4°C along with positive selection of monocytes *via* CD14 magnetic beads ensures purities >90% with minimal signs of artificial gene transcription. These results are in line with a recent study of Lyons et al. demonstrating that positive selection using magnetic microbeads has minimal effect on the activation status of isolated monocytes, lymphocytes or granulocytes [16].

The most critical parameters with regard to artificial gene activation were increased processing temperatures at 20°C and the application of Ficoll for pre-enrichment of PBMCs. These results are in line with another study focussing on the comparison of temperature-dependent cell sorting protocols [10]. Negative selection resulted in lesser purities of only 70%. Regarding the question of how much purity is necessary for gene expression profiling, Szaniszlo et al. presented data that purities above 75% are indistinguishable from a pure sample [18]. When comparing cells isolated by positive or negative sorting, it was obvious that most of the differentially expressed transcripts were contributed by contaminating cells.

Figure 3.
This figure shows the final experimental setup to fractionate peripheral blood in its major cellular constituents, such as granulocytes, monocytes, NK cells, B-, CD4- and CD8-lymphocytes, by the application of magnetic and 4-channel high-speed FACS-sorting. Cells of different rheumatic diseases and healthy controls have been analysed by Affymetrix global gene expression arrays. The upper part of the table shows signal intensities for lineage-specific

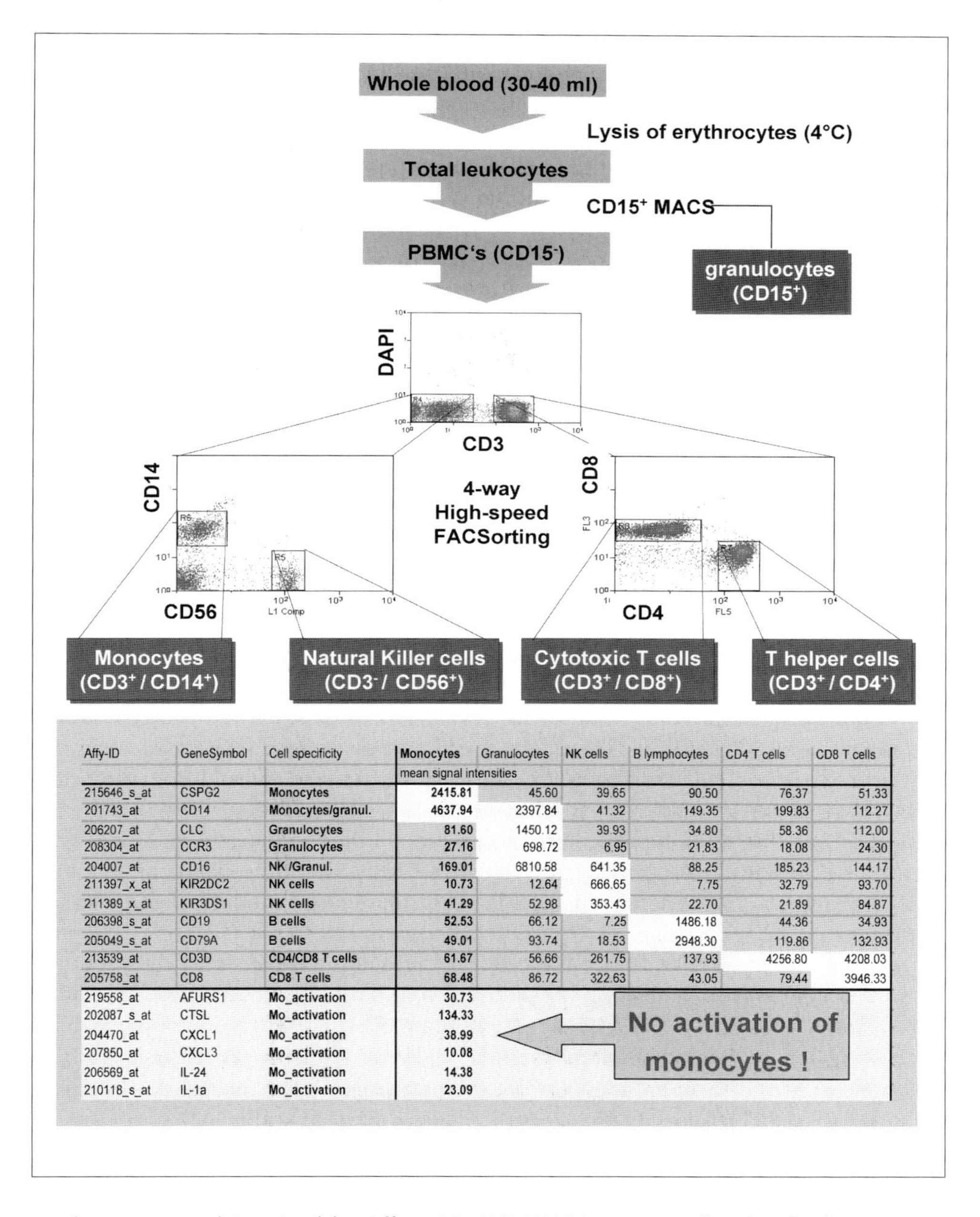

Affy-ID	GeneSymbol	Cell specificity	Monocytes	Granulocytes	NK cells	B lymphocytes	CD4 T cells	CD8 T cells
			mean signal intensities					
215646_s_at	CSPG2	**Monocytes**	**2415.81**	45.60	39.65	90.50	76.37	51.33
201743_at	CD14	**Monocytes/granul.**	**4637.94**	2397.84	41.32	149.35	199.83	112.27
206207_at	CLC	**Granulocytes**	**81.60**	1450.12	39.93	34.80	58.36	112.00
208304_at	CCR3	**Granulocytes**	**27.16**	698.72	6.95	21.83	18.08	24.30
204007_at	CD16	**NK /Granul.**	**169.01**	6810.58	641.35	88.25	185.23	144.17
211397_x_at	KIR2DC2	**NK cells**	**10.73**	12.64	666.65	7.75	32.79	93.70
211389_x_at	KIR3DS1	**NK cells**	**41.29**	52.98	353.43	22.70	21.89	84.87
206398_s_at	CD19	**B cells**	**52.53**	66.12	7.25	1486.18	44.36	34.93
205049_s_at	CD79A	**B cells**	**49.01**	93.74	18.53	2948.30	119.86	132.93
213539_at	CD3D	**CD4/CD8 T cells**	**61.67**	56.66	261.75	137.93	4256.80	4208.03
205758_at	CD8	**CD8 T cells**	**68.48**	86.72	322.63	43.05	79.44	3946.33
219558_at	AFURS1	**Mo_activation**	**30.73**					
202087_s_at	CTSL	**Mo_activation**	**134.33**					
204470_at	CXCL1	**Mo_activation**	**38.99**					
207850_at	CXCL3	**Mo_activation**	**10.08**					
206569_at	IL-24	**Mo_activation**	**14.38**					
210118_s_at	IL-1a	**Mo_activation**	**23.09**					

marker genes as determined by Affymetrix HG-U133A arrays and strikingly demonstrates purities of isolated cells. These results are in accordance with purities determined by FACS analysis. In the lower part transcripts are summarised, known to be differentially regulated in activated monocytes, but which are almost absent in cells after sorting.

High-speed FACS-sorting did not induce significant transcriptional alteration as long as continuous cooling was ensured during the entire procedure. Usually, a sorting speed of 25,000 cells per second and a system pressure of 35 PSI were applied, allowing for the isolation of approximately 2–4 mio. monocytes within 30 min. starting from 100 mio. PBMCs. In principal, the MACS procedure is also applicable to whole blood if only limited numbers of cells are necessary or if amplified mRNA is used. But it must be considered that higher amounts of magnetic beads and specially adapted reagents for whole blood applications will be needed. Problems may arise when preparing blood samples with increased granulocyte numbers (>70%). This will usually not result in highly purified monocyte preparations and, therefore, removing granulocytes using CD15-microbeads is strongly recommended.

In summary, we successfully applied a sequential combination of hypo-osmotic lysis, magnetic and four-channel high-speed FACS-sorting to dissect whole blood samples into their major constituents, such as granulocytes, monocytes, NK cells, B, CD4- and CD8-lymphocytes for their subsequent analysis by global gene expression profiling (Fig. 3). The table included in Figure 3 shows signal intensities for lineage-specific marker genes as determined by Affymetrix HG-U133A arrays and strikingly demonstrates purities of isolated cells with minimal indications for an artificial activation of monocytes. We have successfully applied this sorting strategy in the field of rheumatological research, especially to identify cell-specific gene expression patterns for the classification of chronic-inflammatory rheumatic diseases [19, 20].

In terms of outlook, the ideal, integrated approach to assess cell-specific gene expression profiles should enable the input of whole blood samples at one end and the output of meaningful gene signatures at the other. A first attempt to be aimed in this direction is provided by Sethu et al. [21]. They developed a microfluidic device that achieved leukocyte enrichment without artificial cell activation during processing. In the future, appropriate bioinformatic algorithms may be available to dissect complex gene signatures into their single constituents *in silico*, as discussed by Haeupl et al. [19]. If so, physical cell separation of whole blood could possibly be completely abandoned.

Acknowledgements

Supported by National funding: German Federal Ministry of Education and Research (BMBF) through the National Genome Research Network (Infection & Inflammation Network SIPAGE) and EU-fundings: AUTOROME; LSHM-CT-2004-005264 and AutoCure; LSHB-CT-2006-018661.

References

1 Grant GR, Manduchi E, Pizarro A, Stoeckert CJ, Jr. Maintaining data integrity in microarray data management. *Biotechnol Bioeng* 2003; 84(7): 795–800
2 Wilkes T, Laux H, Foy CA. Microarray data quality – review of current developments. *OMICS* 2007; 11(1): 1–13
3 Brazma A, Hingamp P, Quackenbush J, Sherlock G, Spellman P, Stoeckert C et al. Minimum information about a microarray experiment (MIAME)-toward standards for microarray data. *Nat Genet* 2001; 29(4): 365–71
4 Geschwind DH. Sharing gene expression data: an array of options. *Nat Rev Neurosci* 2001; 2(6): 435–8
5 Imbeaud S, Auffray C. 'The 39 steps' in gene expression profiling: critical issues and proposed best practices for microarray experiments. *Drug Discov Today* 2005; 10(17): 1175–82
6 Shi L, Reid LH, Jones WD, Shippy R, Warrington JA, Baker SC et al. The MicroArray Quality Control (MAQC) project shows inter- and intraplatform reproducibility of gene expression measurements. *Nat Biotechnol* 2006; 24(9): 1151–61
7 Bakay M, Chen YW, Borup R, Zhao P, Nagaraju K, Hoffman EP. Sources of variability and effect of experimental approach on expression profiling data interpretation. *BMC Bioinformatics* 2002; 3: 4
8 Spruill SE, Lu J, Hardy S, Weir B. Assessing sources of variability in microarray gene expression data. *Biotechniques* 2002; 33(4): 916–20
9 Whitney AR, Diehn M, Popper SJ, Alizadeh AA, Boldrick JC, Relman DA et al. Individuality and variation in gene expression patterns in human blood. *Proc Natl Acad Sci USA* 2003; 100(4): 1896–901
10 Debey S, Schoenbeck U, Hellmich M, Gathof BS, Pillai R, Zander T et al. Comparison of different isolation techniques prior gene expression profiling of blood derived cells: impact on physiological responses, on overall expression and the role of different cell types. *Pharmacogenomics* J 2004; 4(3): 193–207
11 Goronzy JJ, Weyand CM. Rheumatoid arthritis. *Immunol Rev* 2005; 204: 55–73
12 Hoffman, E.P. Expression profiling – best practices for data generation and interpretation in clinical trials. *Nat Rev Genet* 2004; 5(3): 229–37
13 Han ES, Wu Y, McCarter R, Nelson JF, Richardson A, Hilsenbeck SG. Reproducibility, sources of variability, pooling, and sample size: important considerations for the design of high-density oligonucleotide array experiments. *J Gerontol A Biol Sci Med Sci* 2004; 59(4): 306–15
14 Macgregor S. Most pooling variation in array-based DNA pooling is attributable to array error rather than pool construction error. *Eur J Hum Genet* 2007; 15(4): 501–4
15 Viale A, Li J, Tiesman J, Hester S, Massimi A, Griffin C et al. Big results from small samples: evaluation of amplification protocols for gene expression profiling. *J Biomol Tech* 2007; 18(3): 150–61
16 Lyons PA, Koukoulaki M, Hatton A, Doggett K, Woffendin HB, Chaudhry AN et al.

Microarray analysis of human leucocyte subsets: the advantages of positive selection and rapid purification. *BMC Genomics* 2007; 8: 64

17 Galbraith DW, Elumalai R, Gong FC. Integrative flow cytometric and microarray approaches for use in transcriptional profiling. *Methods Mol Biol* 2004; 263: 259–80

18 Szaniszlo P, Wang N, Sinha M, Reece LM, Van Hook JW, Luxon BA et al. Getting the right cells to the array: Gene expression microarray analysis of cell mixtures and sorted cells. *Cytometry A* 2004; 59(2): 191–202

19 Haeupl T, Gruetzkau A, Gruen J, Radbruch A, Burmetser GR. Expression analysis of rheumatic diseases, prospects and problems. In: Holmdahl R (ed): *The Hereditary Basis of Rheumatic Diseases*. Basel, Boston, Berlin: Birkhäuser; 2007; 119–30

20 Mahr S, Burmester GR, Hilke D, Gobel U, Grutzkau A, Haupl T et al.. Cis- and trans-acting gene regulation is associated with osteoarthritis. *Am J Hum Genet* 2006; 78(5): 793–803

21 Sethu P, Moldawer LL, Mindrinos MN, Scumpia PO, Tannahill CL, Wilhelmy J et al. Microfluidic isolation of leukocytes from whole blood for phenotype and gene expression analysis. *Anal Chem* 2006; 78(15): 5453–61

Sample preparation

Skin and skin models

Olaf Holtkötter and Dirk Petersohn

Phenion GmbH & Co. KG, Merowingerplatz 1a, 40225 Düsseldorf, Germany

Abstract

Skin inflammation is an often-occurring phenomenon in a wide range of skin pathologies. In order to understand the basics of the underlying molecular mechanisms, early responses such as gene expression changes in this tissue are the focus of numerous studies. The prerequisite for gene expression analysis is the isolation of relevant tissue samples and the subsequent preparation of high quality RNA.

Different sources of tissue samples (e.g. skin biopsies, suction blister, epidermis models and bio-engineered full thickness skin), their benefits and disadvantages are discussed. Furthermore, we briefly describe ways of isolating RNA and assessing its quality, which is essential for performing successful gene expression studies.

Introduction

Healthy human skin forms a very efficient barrier to the environment, thus protecting the organism from a multitude of threats such as excessive water loss, pathogenic germs, extreme temperatures and so forth. As the largest organ of the human body, skin covers a surface of approx. 2.0 m^2 and constitutes roughly 12% of body weight, whereas the thickness of skin is largely dependent on its localization. It ranges from 0.5 mm on, e.g. the eye-lid, to several millimeters on the palms and soles.

Skin is composed of two major compartments, the outer *epidermis* and the underlying *dermis* (Fig. 1). Both parts are specialized regarding their functional properties, physiology and architecture.

The epidermal architecture is typified as stratified epithelium and harbors four distinguishable layers that change in morphology, as the differentiation status of the constituting cells change. The *stratum basale* is composed of a single layer of basal keratinocytes that are attached to the basement membrane. These continuously dividing cells give rise to a more differentiated epidermal layer, the *stratum spinosum*, with its cells' typical prickly appearance. On their way to the skin surface

Microarrays in Inflammation, edited by Andreas Bosio and Bernhard Gerstmayer

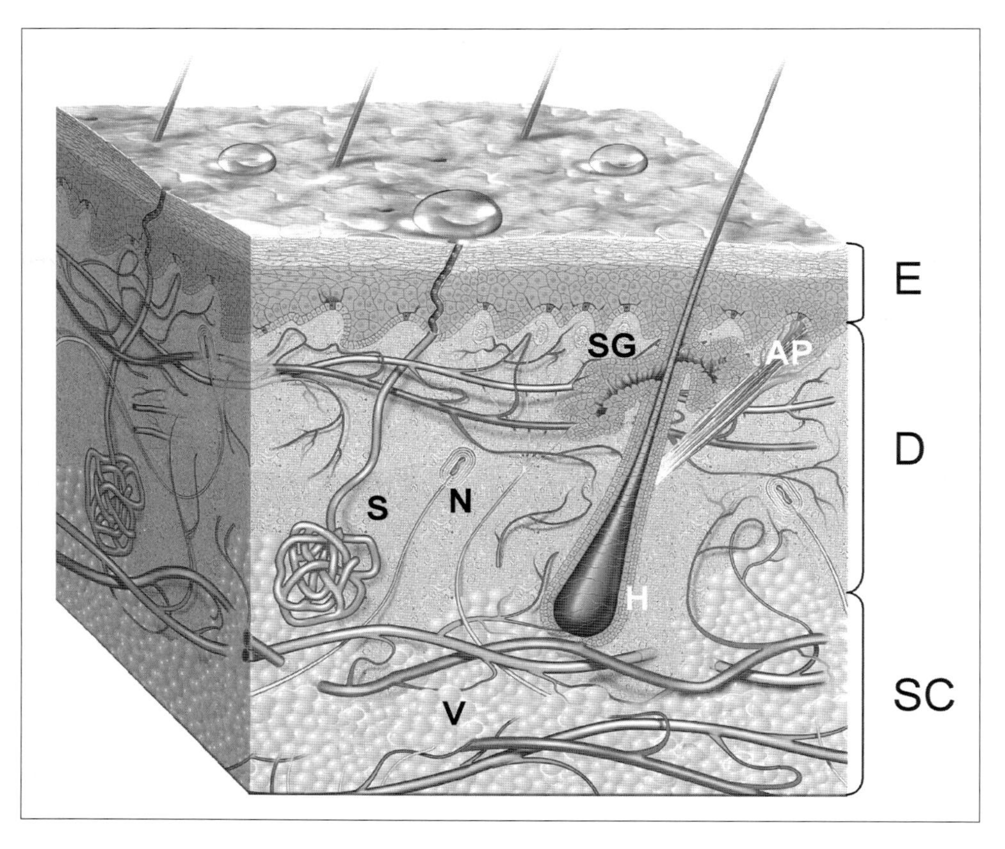

Figure 1
Organization of the skin. The skin is composed of two major compartments, the epidermis and the underlying dermis, in which different appendages are embedded. E: epidermis, D: dermis, SC: subcutis, H: hair, V: vessels, S: sweat gland, SG: sebaceous gland, N: nerve, AR: arrector pili *muscle*

the cells differentiate further, start to flatten, and accumulate granules containing basophilic keratohyalin and lamellar bodies, a typical feature of the *stratum granulosum*. In the final layer, the *stratum corneum*, the anucleate corneocytes arrange in a brick-and-mortar like structure in concert with the lipids from lamellar bodies.

Subjacent to the epidermis, the *dermis* functions as a physiological support and imparts elasticity, firmness, tensile strength etc. The dermis is dense connective tissue that consists of cells – in the majority dermal fibroblasts –, a fibrous meshwork made up of glycoproteins and glycosaminoglycanes and further specialized structures. Additionally, different types of appendages are present in skin. The pilosebaceous unit, e.g. consists of the hair shaft and follicle, the sebaceous gland, and the *arrector pili* muscle. With a few exemptions pilosebaceous units are distributed all over the body, most prominently in the human scalp. The sebaceous glands are tightly linked to the hair follicle and secrete sebum, an oily substance composed of triglycerides, squalene, wax esters and free fatty acids [1, 2]. The pilosebaceous unit is also the location where different types of *Acne* – most prominent *Acne vulgaris* – are located and lead to skin inflammation [3].

Beside the above mentioned major cellular components, further specialized cells are interspersed in the skin. Melanocytes, e.g. are located in the *stratum basale* and are responsible for the production of pigment-containing granules that, in turn, impart skin color [4]. The immunological competence of the skin, in addition to the keratinocyte's contribution, is especially exerted by Langerhans cells, which are bone marrow derived dendritic antigen-presenting cells, and mast cells, a leukocyte subset that takes part, e.g. in the inflammation process during wound repair [5].

The three-dimensional structure of skin, as is described above, can be reconstructed as organotypical models, which were initially developed for grafting purposes and have since gained an increasing role in biomedical research (Fig. 2). Starting with the well-known and simple collagen gel model introduced by Bell et al. [6], which represents an artificial dermis, the complexity of the models has increased, nowadays comprising reconstituted epidermis, full thickness models with a dermal compartment and an epidermis, models containing endothelial cells, dendritic cells or melanocytes, and so forth. An increasing number of models are now commercially available. The striking advantage of skin models is certainly their high degree of uniformity and their ease of handling, which leads to standardized experiments and very reproducible results. This is demonstrated, e.g. in the ambition to use skin models for human safety assessments, evaluating the irritant potential of chemicals by testing them on reconstructed organotypic skin [7].

To analyze molecular events in skin cells we here review state-of-the-art techniques to prepare RNA from skin tissues for further downstream manipulation. We focus on skin tissue from humans or tissue-engineered human skin, some minor technological adoptions, however, may be necessary when analyzing corresponding tissues from animals.

Sources of skin RNA

In vivo skin

The starting point for RNA preparation is the isolation and protection of the desired tissue, which should be performed as quickly as possible to avoid degradation of RNA.

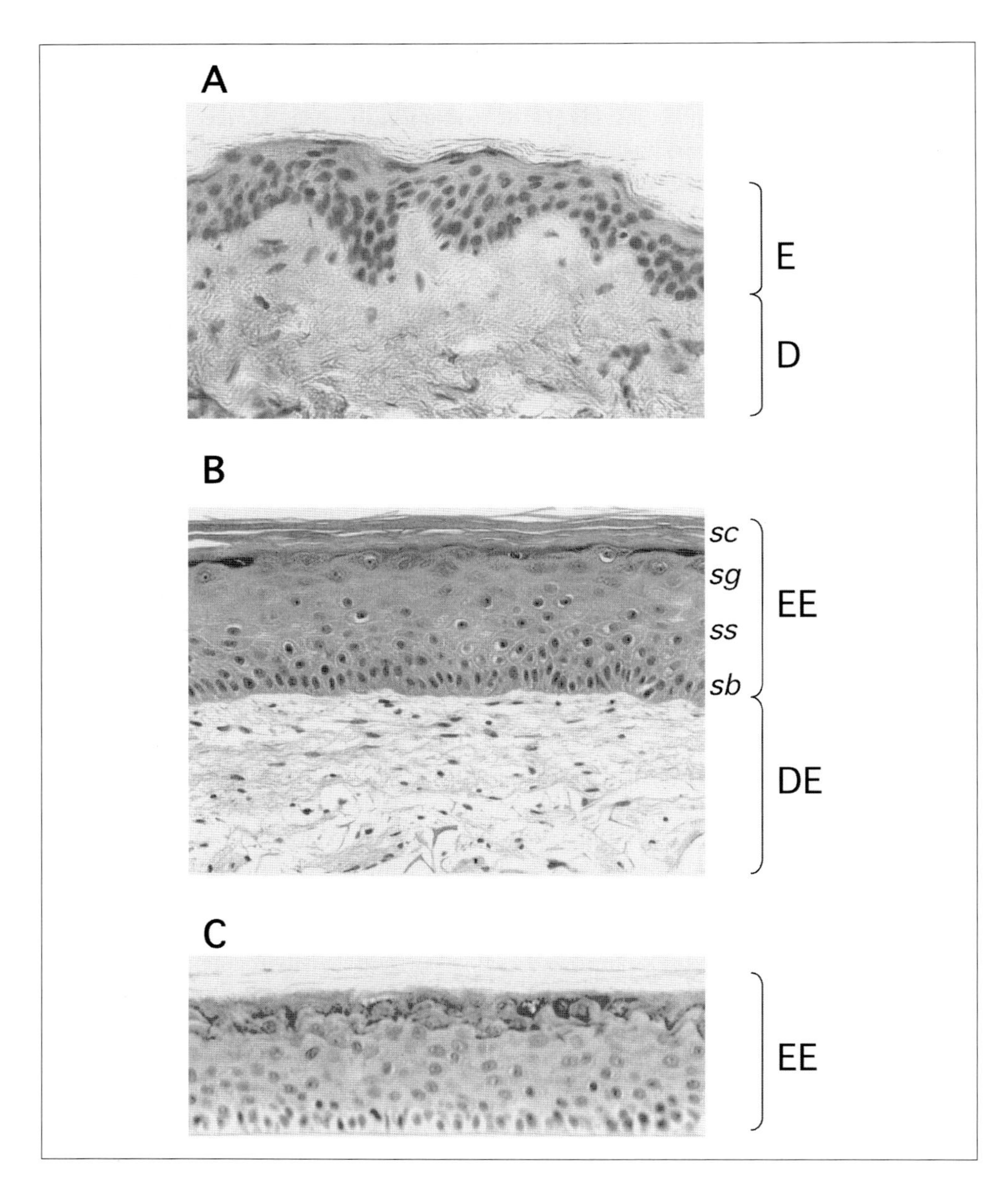

Figure 2.
Histological sections of skin and skin models. Sections were taken from paraffin-embedded tissues and stained with hematoxylin-eosin. A) skin from mamma reduction (E: epidermis, D: dermis), B) Phenion® Full Thickness Skin Model (EE: epidermal equivalent, DE: dermal equivalent, sb: stratum basale, *ss:* stratum spinosum, *sg:* stratum granulosum, *sc:* stratum corneum*) [20], C) Phenion's epidermis model (EE: epidermal equivalent).*

Lab-animals, e.g. mice, are regularly used in biomedical research, thus avoiding inter-individual differences, which in turn is a typical hurdle when working with human tissues. For the isolation of skin from lab animals, they are either killed before skin processing, or set under general anesthesia to excise a defined skin area. The animal's fur should be removed quickly with a pair of scissors or a razor blade before skin excision, to avoid any disturbance in the following RNA isolation.

Methods to obtain skin material from humans can be distinguished by the invasiveness of the method and, hence, by the layers the isolated skin comprises. When RNA from all layers of the skin is desired, usually a punch biopsy is taken, which can be performed under local anesthesia. Because of the average size of the biopsies, the tissue can be directly shock-frozen or stored in stabilization solutions without RNA degradation. Punch biopsies with a diameter of 4 mm (50 mg) from buttock skin typically results in about 20–30 µg total RNA. Beside punch biopsies, skin tissue often originates from surgery of skin pathologies, such as tumor excisions etc., or from plastic surgery, e.g. mamma reductions or face lifts. Waste material from plastic surgery in particular offers an alternative, if healthy and untreated skin is needed in larger quantities.

Depending on the type and site of surgery, large skin pieces must be reduced in size to enable quick freezing or protection in a stabilizing solution. This can either be done by applying biopsy instruments (e.g. with a diameter of 10 mm) on the excised material, or by using scalpels or scissors. A drawback of this method is the prolonged period of time in which RNases prove effective.

The RNA yield from excised skin tends to depend on the site of excision, and on the individual from whom the skin is taken. As a rule of thumb, human skin yields about 0.2–0.4 µg total RNA per milligram tissue, and this suffices to estimate the amount of RNA that can be isolated from excision sites as different as temple, cheek, arm, breast, and buttock.

In cases where only epidermal RNA is required, the skin's epidermis can be isolated from a suction blister. This technique uses a suction device to apply a vacuum to the skin, resulting in the separation of the epidermis from the dermis, and the generation of a skin blister. The epidermal sheet is removed from the blister to prepare the RNA. Hochberg et al. [8] used the method of suction blistering to investigate effects of phototherapy on psoriatic skin by microarray analysis.

A very interesting approach that circumvents the invasiveness of all before-mentioned methods is tape-stripping of the skin. Several pieces of sticky tape are placed consecutively on a certain area of the skin, removed and collected. Cells from the living layers of the epidermis are isolated with progressive tape-stripping. Wong et al. [9, 10] describe the application of this method for RNA isolation and downstream applications.

Finally, it should be mentioned that skin biopsies can be cultured *ex vivo* for a limited time. In this regard the handling of the tissue is similar to the handling of skin models, which is described in the following paragraph.

Skin models

Harvesting and storage of skin models for RNA preparation is very comfortable from an experimental point of view because these tissue pieces are of suitable size for direct shock-freezing, even though sometimes, depending on the kind of model, folding before freezing may be necessary. In general, the handling of skin models is rather reliable, but they tend to react to mechanical stress and therefore should be transported and handled gently. In addition, as with monolayer cell cultures, attention should be paid to keeping the time between taking the models out of the incubator and freezing them short, since the environmental change induces relevant adaptations in the gene expression profile of the models. However, for a trained person it is even possible to cut the models into halves, using one half for RNA preparation and the other half for histology or protein isolation [11]. The typical total RNA yield from epidermis models, e.g. those commercially available from SkinEthic, is about 20–30 μg/model, full thickness models, e.g. from Phenion or MatTek, lead to 40–60 μg/model. In our experience, the RNA isolated from full thickness models derives about a third of the dermal equivalent and roughly two-thirds of the epidermal compartment. A massive decrease in the RNA yield is typically a good indicator of a high degree of damage in the model, most probably caused by the selected treatment.

Splitting of dermis and epidermis

Splitting of dermis and epidermis provides the opportunity to analyze epidermal RNA from full thickness biopsies or skin models. Separating skin into epidermis and dermis before isolating RNA still remains a challenge, since general protocols include a prolonged incubation step with the splitting reagent. This incubation step can affect the gene expression profile of the tissue, leading to an additional bias in the following analysis of the RNA.

Different methods for epidermis-dermis splitting are described in the literature, comprising amongst others enzymatic digestion and chemical separation. A comparison of these methods in terms of RNA quality after splitting of skin biopsies is described by Trost et al. [12] who recommend the use of ammonium thiocyanate for subsequent gene expression studies. In principle, the methods used for skin biopsies can also be applied to full thickness skin equivalents, with the method and exact protocol requiring optimization depending on the type of model. However, since the dermal-epidermal connection is normally somewhat weaker in reconstituted skin, these protocols are, in general, more gentle because of their shorter incubation times, so that the expected stress on the models can be assumed to be lower than in skin biopsies.

When working with full thickness skin models based on a stabilized matrix [13, 14] incubation with a solution of 10 mM EDTA for 20 minutes at 37°C is already sufficient to separate the epidermis equivalent from the dermal matrix of the model, and the splitting can be improved by using 0.5 mg/ ml thermolyase [15]. Nevertheless, all of these treatments have an impact on the gene expression profile and precaution should be taken when interpreting results from RNA preparations thus isolated.

Among others, a new and promising alternative to the above mentioned methods is laser microdissection [16, 17] which offers the opportunity to isolate defined groups of cells or even single cells from a tissue. Since tissue sections are a prerequisite for applying this technology, the challenge of this procedure is the preparation of sections under conditions that reduce RNA degradation as far as possible.

Preparation of total RNA from skin

The prerequisite for a successful RNA preparation is the consideration of standards in RNA handling, as described in classical protocol books such as *Molecular Cloning. A Laboratory Manual* [18] and *Current Protocols in Molecular Biology* [19]. The contamination of samples with RNases is still the major source of failure in RNA preparations and can be easily avoided by a well-trained person and proper preparation of lab material.

The crucial step in RNA preparation from skin is solubilization, since skin is a rather robust tissue, which means that high mechanical forces must be applied for its homogenization. For a smaller amount of samples this can be accomplished with, e.g. an Ultrathurax. As soon as the number of samples increases, the application of bead mills, in which several samples can be homogenized simultaneously, is recommended. Apart from the restrictions related to tissue disruption, any of the different standard protocols for RNA isolation can be used, i.e. protocols employing organic solvents such as the traditional guanidium isothiocyanate/phenol protocol or protocols based on solid phase purification in (spin) columns. Although the high content of lipophilic molecules from the skin's lipid barrier and charged dermal mucopolysaccharides might interfere with RNA isolation to a certain extent, the different methods will all attain a reasonable yield of RNA, provided that the sample size is of the magnitude of a 10 mm punch biopsy. It is needless to mention that protocols must be optimized when the sample size decreases or the yield or quality of the preparation is to be improved.

Our standard protocol for total RNA preparation from skin biopsies and skin models is based on Quiagen's RNeasy Protocols. The tissue (up to 150 mg) is placed in a guanidine-isothiocyanate containing lysis buffer and homogenized with a Retsch Mixer Mill MM301, using 5 mm steel beads. Skin models are ground for five minutes, *ex vivo* skin for 10 minutes at 30 Hz. The homogenate is digested

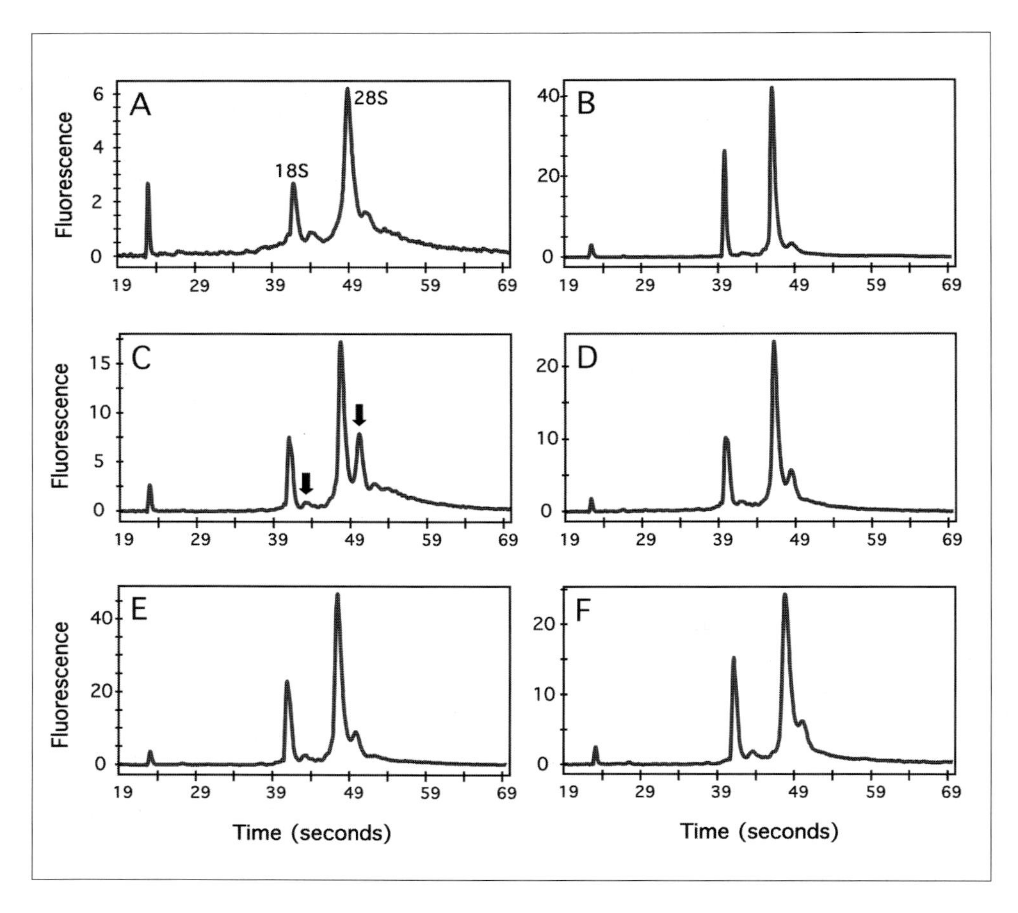

Figure 3
Electropherogramms of RNA preparations from different 'skin' sources. A) skin from mamma reduction, B) keratinocytes, C) reconstituted epidermis, D) epidermis from splitting of a full thickness model, E) Phenion's full thickness model [20], F) full thickness model according to Black et al. [13]. Depending on the source of RNA, the peaks following the 18 S- and 28 S-peaks show typical differences (arrows).

with proteinase K, freed from the debris and further processed using spin columns. Normally, we do not find any quality loss in our preparations when working with 12 samples in parallel.

The quality of the RNA preparation clearly impacts the results of downstream applications and should therefore be assessed. Traditionally, the integrity and yield of a total RNA sample is controlled by spectrophotometry and denaturing agarose gel electrophoresis. These methods suffer from the high amount of total RNA needed and, in the latter case, from the rather low sensitivity. In recent years, the quality assessment with the lab-on-a-chip technology of Agilent's Bioanalyzer 2100 has evolved more and more as a state-of-the-art technique, and similar devices from other companies, e.g. Biorad's Experion, are also emerging. Especially when RNA preparation is a routine method in the lab, such devices are highly recommended for quality control, since many more details are made visible with this technology. Figure 3 shows the electropherogramms of different samples, indicating typical differences depending on the tissue's origin.

RNA preparations that are of high quality can be stored at –20°C even for years, and will survive transportation at room temperature without loss of quality. However, long term storage at –80°C is recommended. The preparation can be used for any downstream application, e.g. mRNA enrichment, linear amplification and, of course, microarray analysis.

Acknowledgements
We would like to thank Waltraud Knieps-Massong, Anke Kolbe and Günter Schmidt, for their technical assistance in many of the experiments described in this chapter. Some of the work was done as part of collaboration with colleagues from Miltenyi, and we are grateful for many fruitful discussions.

References

1 Thody AJ, Shuster S (1989) Control and function of sebaceous glands. *Physiol Rev* 69: 383–416
2 Steward ME, Downing DT (1991) Chemistry and function of mammalian sebaceous lipids. In: PM Elias (ed): *Advances in Lipid Research*, Vol. 24, Skin Lipids, San Diego Academic Press, San Diego, 263–302
3 Zouboulis CC, Eady A, Philpott M, Goldsmith LA, Orfanos C, Cunliffe WC, Rosenfield R (2005) What is the pathogenesis of acne? *Exp Dermatol* 14: 143–52
4 Lin JY, Fisher DE (2007) Melanocyte biology and skin pigmentation. *Nature* 445: 843–50
5 Schröder JM, Reich K, Kabashima K, Liu FT, Romani N, Metz M, Kerstan A, Lee PH, Loser K, Schön MP, Maurer M, Stoitzner P, Beissert S, Tokura Y, Gallo RL (2006) Who is really in control of skin immunity under physiological circumstances – lymphocytes, dendritic cells or keratinocytes? *Exp Dermatol* 15: 913–29
6 Bell E, Ivarsson B, Merrill C (1979) Production of a tissue-like structure by contraction of collagen lattices by human fibroblasts of different proliferative potential *in vitro*. *Proc Natl Acad Sci USA* 76: 1274–8
7 Welss T, Basketter DA, Schröder KR (2004) *In vitro* skin irritation: facts and future. State of the art review of mechanisms and models. *Toxicology in Vitro* 8: 231–243

8 Hochberg M, Zeligson S, Amariglio N, Rechavi G, Ingber A, Enk CD (2007) Genomic-scale analysis of psoriatic skin reveals differentially expressed insulin-like growth factor-binding protein-7 after phototherapy. *Br J Dermatol* 156: 289–300

9 Wong R, Tran V, Morhenn V, Hung SP, Andersen B, Ito E, Wesley Hatfield G, Benson NR (2004) Use of RT-PCR and DNA microarrays to characterize RNA recovered by non-invasive tape harvesting of normal and inflamed skin. *J Invest Dermatol* 123: 159–67

10 Wong R, Tran V, Talwalker S, Benson NR (2006) Analysis of RNA recovery and gene expression in the epidermis using non-invasive tape stripping. *J Dermatol Sci* 44: 81–92

11 Welss T, Matthies W, Schroeder KR (2007) Compatibility testing *in vitro*: a comparison with *in vivo* patch test data. *Int J Cosm Sci* 29: 143

12 Trost A, Bauer JW, Lanschutzer C, Laimer M, Emberger M, Hintner H, Onder K (2007) Rapid, high-quality and epidermal-specific isolation of RNA from human skin. *Exp Dermatol* 16: 185–90

13 Black AF, Bouez C, Perrier E, Schlotmann K, Chapuis F, Damour O (2005) Optimization and characterization of an engineered human skin equivalent. *Tissue Eng* 11: 723–33

14 Shahabeddin L, Berthod F, Damour O, Collombel C (1990) Characterization of skin reconstructed on a chitosan-cross-linked collagen-glycosaminoglycan matrix. *Skin Pharmacol* 3: 107–14

15 Walzer C, Benathan M, Frenk E (1989) Thermolysin treatment: a new method for dermo-epidermal separation. *J Invest Dermatol* 92: 78–81

16 Becker B, Roesch A, Hafner C, Stolz W, Dugas M, Landthaler M, Vogt T (2004) Discrimination of melanocytic tumors by cDNA array hybridization of tissues prepared by laser pressure catapulting. *J Invest Dermatol* 122: 361–8

17 Nanney LB, Caldwell RL, Pollins AC, Cardwell NL, Opalenik SR, Davidson JM (2006) Novel approaches for understanding the mechanisms of wound repair. *J Investig Dermatol Symp Proc* 11: 132–9

18 Sambrook J, Fritsch EF, Maniatis T (eds) (1989) *Molecular Cloning. A Laboratory Manual.* Cold Spring Harbor Laboratory Press, Cold Spring Harbor

19 Ausubel FM, Brent R, Kingston RE, Moore DD, Seidman JG, Struhl K, Smith JA (eds) (2007) *Current Protocols in Molecular Biology.* John Wiley & Son

20 Mewes KR, Raus M, Bernd A, Zoller NN, Sattler A, Graf R (2007) Elastin expression in a newly developed full-thickness skin equivalent. *Skin Pharmacol Physiol* 20: 85–95

Sample preparation

Adipose tissue

Nathalie Viguerie

Inserm, U858, Laboratoire de Recherches sur les Obésités, Institut de Médecine Moléculaire de Rangueil, Toulouse, 31432 France; Université Paul Sabatier, Institut Louis Bugnard IFR31, Toulouse, 31432 France; Centre Hospitalier Universitaire de Toulouse, Toulouse 31059, France

Abstract

Despite inherent difficulties, the use of technologies in Omic such as DNA microarrays, that allow a comparison of global expression changes in thousands of genes between different conditions, appears to be a useful tool to study the influence of nutrition and gene-environment interactions in complex pathologies such as obesity, and thus to advance research.

The purpose of the present review is to provide some advice for fat specimen collection and RNA preparation in view of gene expression profiling analysis. Our data indicate that the use of RNA-stabilization reagents is not necessary. Isolated or cultured cells should be stored frozen in lysis buffer. Isolated adipose cells or pieces of adipose tissue must be shock-frozen in liquid nitrogen before storage. Total RNA extraction can then be performed using optimized tissue lipid RNA extraction systems. For clinical projects, fat biopsies of 0.5 g are recommended. Total RNA should be extracted using a modified tissue lipid extraction kit. Finally, some selected examples will show how previous transcriptomic approaches to human adipose tissue biopsies have contributed to advancing knowledge in the field of obesity, as much at the cellular as at the tissue level.

Introduction

The identification of physiological and biological factors that underlie the metabolic disturbances observed in obesity is a key step to providing better therapeutic outcomes. The development of obesity is an ongoing process extending over years that results from inappropriate adaptation of energy balance control to either a primarily increased energy intake or reduced energy expenditure. This leads to an accumulation of surplus energy in the form of triglyceride in adipose tissue. At present, this tissue is no longer considered a passive energy keeper but rather a critical component in the regulation of energy balance, due to its intense metabolic activity and its ability to produce various molecules called adipokines. In the complex picture of the pathophysiology of obesity, adipose tissue gene expression profiling as well as other

Microarrays in Inflammation, edited by Andreas Bosio and Bernhard Gerstmayer

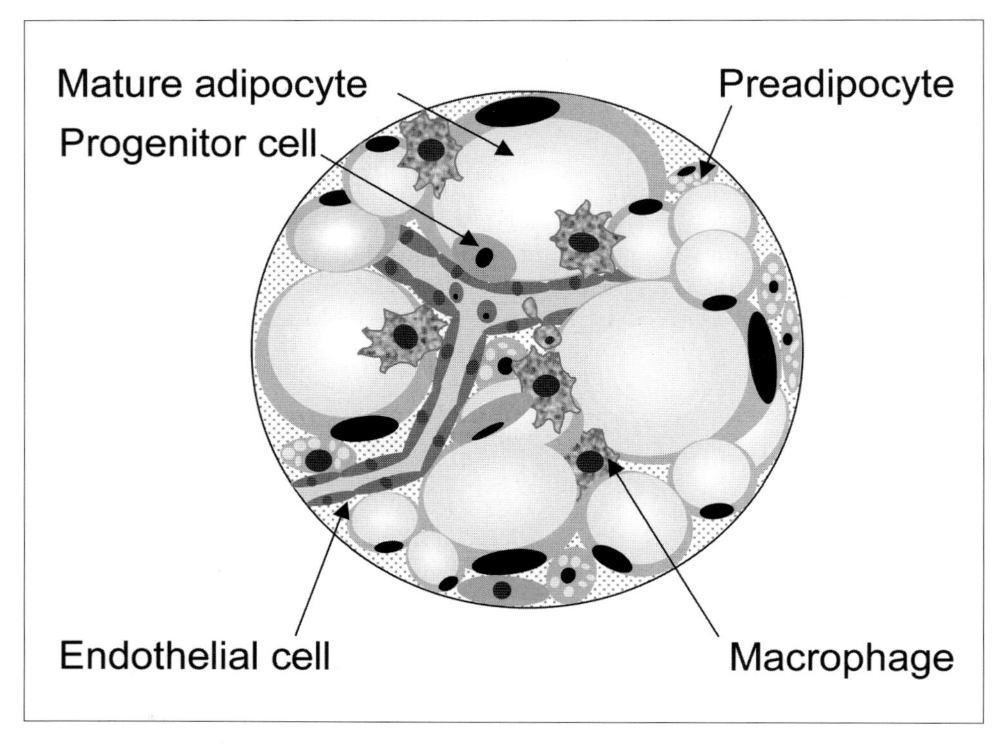

Figure 1.
Cartoon of a section of adipose tissue from obese subjects.
Fat tissue is composed of several cell types: mature adipocytes and various other small cells such as preadipocytes, progenitor cells, endothelial cells, and macrophages, usually referred to as the "stroma vascular fraction".

Omic approaches (proteomics, metabolomics) may contribute to revealing the role of some signals and could then provide a better understanding of the mechanisms of energy homeostasis. For *in vivo* gene expression studies in humans, key peripheral tissues (mainly adipose tissues, skeletal muscle, and to a lesser extent heart and liver) are accessible by biopsy during clinical investigation protocols. However, despite outstanding cell size when compared to other cells, the presence of large amounts of intracellular triglycerides within adipocytes makes the amount of RNA available from adipose tissue very low (Fig. 1). Alternatively, primary cultures of preadipocytes isolated from the stroma vascular fraction (SVF) or established adipose cell lines can serve as *in vitro* models.

The transcriptomic strategy is now the most widely used Omic strategy in the field of obesity research, due to its capacity to simultaneously quantify the entire transcript population in a single experiment [1]. The recent multicentric European study MolPAGE (Molecular Genotyping to Accelerate Genomic Epidemiology) aims to tackle diabetes and vascular diseases through the development and application of a range of Omic technology platforms, to carry out "molecular phenotyping" on a medium to epidemiological scale (www.molpage.org). In this program our contribution is to develop robust, optimised protocols for sample collection, preparation and storage for RNA biobanked samples of fat biopsies. These RNAs should be suitable for transcriptomics applications, notably the measurement of gene expression (RT-qPCR, DNA microarrays) in large-scale epidemiological studies.

The purpose of the present paper is to review recent optimization in adipose tissue sampling and total RNA extraction for gene expression profiling experiments. It will also emphasize how the application of DNA microarray technique on human adipose tissue has refined our understanding of currently known molecules as well as identified novel adipokines which are involved in energy homeostasis and also in other unknown pathways.

Specimen collection

Once cultured cells or adipose tissue is harvested, its RNA becomes extremely unstable. The processing of adipose samples has thus been optimized with respect to the total RNA quality and yield.

Cell culture

Culture medium is removed and cultured cells are rinsed with sterile phosphate buffer saline (PBS) at room temperature, disrupted in high-salt lysis buffer then immediately stored at –80°C prior total RNA preparation.

Isolated cells

Collagenase digestion of adipose tissue results in the separation of mature adipose cells from the stroma vascular fraction (SVF) that contains preadipocytes, macrophages, endothelial cells and other cell types (Fig. 1). The tissue is digested using 0.5 mg/ml collagenase in Krebs Ringer bicarbonate buffer containing 3.5% (w/v) bovine serum albumin at 37°C under gentle agitation. After filtering through a 150–200 µm nylon mesh the infranatant is separated from the floating mature adipocytes, which are washed twice with sterile PBS before storage at –80°C. The infranatant containing the non fatty cells is centrifuged at 800 g for 10 minutes at 4°C, washed twice with sterile PBS and crushed in lysis buffer before storage at –80°C. Regarding gene expression profiling experiments, attention must be paid to the digestion step since it has been reported that prolonged adipose tissue collagenase digestion may induce major transcriptional changes and, in particular, induce inflammatory mediators [2]. Usually 30–45 min are sufficient for full tissue digestion.

Surgical samples and needle biopsies

Surgical samples (abdominal dermolipectomies, visceral fat) are dissected from vessels and possibly skin. To harvest subcutaneous adipose tissue, a mini liposuction with a 12G needle and Hepafix® syringe (B. Braun) after tissue infiltration of lidocaine 1% is a safe and painless method. This is usually performed in the abdominal region (10 cm laterally from umbilicus). This technique has been widely used in various clinical protocols for harvesting hundreds of subcutaneous adipose tissue biopsies. Usual recovery is about 0.5 to 1 g depending on the physician's expertise and the fat mass of the subject. The adipose tissue is rinsed with sterile saline solution at room temperature, then coagulated blood clots are discarded with a sterile spatula or a needle as quickly as possible. The tissue is then frozen in liquid nitrogen before storage at –80°C. If more tissue is required, a second biopsy may be performed which should be handled and stored in a separate tube.

Sample storage

Several commercial RNA-stabilization reagents with protective effect are available in order to prevent RNA degradation and to improve the quality of microarray gene expression data. However, our experience indicates that for long term storage such preservatives may induce sporadic genomic DNA contamination. Consequently, no tissue preservative is used prior to storage until RNA preparation, and samples are thus processed as swiftly as possible. In such conditions similar total RNA quantity (µg) and quality are obtained as for fresh adipose tissue samples. In the context of the MolPage program, we are currently investigating whether long term storage of fat samples is feasible and has no negative impact on RNA integrity. Briefly, tissues which have been frozen in liquid nitrogen (with or without use of RNA preservatives) can be stored at –80°C up to three years without affecting total RNA yield or quality. Samples already lysed in chaotropic buffer have been stored up to four years without any sign of RNA degradation.

Total RNA extraction

Many technologies and commercial kits are available for the purification of mRNA or total RNA from cells and adipose tissues. Regarding primary cultured cells, isolation of total RNA using standard RNA isolation methods results in an average yield of 8.2 ± 2.4 μg (2–13 μg) total RNA from 25×10^4 cells. However, none of these standard RNA isolation methods works well on triglyceride-enriched samples and the yields obtained are typically <10% of the estimated RNA content (1–10 μg total RNA/g of tissue *vs.* 60 μg total RNA/g of tissue). When compared to a CsCl gradient or phenol-chloroform methods, the use of spin-column purification systems significantly shortens preparation time.

Five single-step methods for total RNA extraction, i.e., RiboPure (Ambion), RNA STAT-60™ (Tel-Test Inc.), NucleoSpin RNA II (MachereyNagel), RNeasy Mini, and RNeasy Lipid Tissue Kits (QIAGEN) were evaluated on human adipose tissue samples. Among the five extraction methods tested, three yielded very few or partly degraded samples and the QIAGEN method yielded sufficiently good quality total RNA, which appeared to be free of contamination with genomic DNA. Despite the recent occurrence of novel kits dedicated to fatty animal tissues, efforts have been made in-house to optimize total RNA extraction from adipose tissue samples of 0.1–1 g in order to prevent the loss of precious samples from patients. By improving the RNA recovery from samples rich in fat it becomes possible to limit the biopsy sample required, thereby facilitating access to adipose tissue for future large-scale epidemiological studies.

An optimized protocol was developed for the total RNA preparation. The present total RNA extraction protocol uses modified phenol/guanidine based methods, silica-gel membrane purification and microspin technology. This produces total RNA which contains messenger, transfer, and ribosomal RNA free of the ribosomal 5S form and genomic DNA. Frozen samples containing lysis buffer (i.e. cultured cells) are thawed on ice. Frozen crude adipose tissue samples are immediately crushed in phenol-containing denaturing solution using a Polytron homogenizer until the sample is uniformly homogeneous. A 20% solution of chloroform (vol/vol) is then added under a ventilated hood and the sample is vigorously vortexed. After centrifugation the upper aqueous phase is mixed with 70% ethanol and transferred onto the microspin column. After the washing steps total RNA is eluted using warm (55°C) RNase free water.

The optimal quantity of adipose tissue for total RNA preparation is about 0.5g (Tab. 1). This corresponds to a mean total RNA yield of 43.93 ± 15.66 μg per g of tissue, and produces enough total RNA for both microarray experiments and RT-qPCR.

The quality of mRNA is a major concern because of its extreme instability with contaminating RNases. This is particularly important when the goal is to analyze gene expression quantitatively, especially in clinical diagnostics. Conventional methods to assess total RNA yield and quality (260/280 nm ratio, ethidium bromide stained agarose gel, spectrophotometer quantitation) are material- and time-consuming and exhibit rather low sensitivity. Therefore, a need exists for easy-to-use assay tools for sensitive quantitation of RNA requiring small sample volumes

Table 1 - Average total RNA yield from fat biopsies with different weight.
Total RNA was extracted using a modified phenol/chloroform based preparation and silica-gel microspin purification. Data are expressed as mean ±SD

Number of samples	Weight of fat biopsies (g)	Total RNA (μg)	Min.-max. amounts (μg)
24	1	34.25±21.97	12.97–96.00
12	0.5	25.18±10.03	11.55–41.46
12	0.2	5.08±0.94	2.75–6.20
18	0.1	2.83±0.52	1.98–3.20
18	0.05	1.10±0.30	0.59–1.48

(<3 μl) for analysis. The Nanodrop® Technologies spectrophotometer is based on sample retention technology coupled to a full-spectrum UV/Vis spectrophotometer to assess nucleic acid concentration using 1 μl. The ability to perform laboratory operations on a small scale using miniaturized (lab-on-a-chip) devices is also very appealing. This technology utilizes a network of channels and wells that are etched on glass or polymer chips to build mini-labs. This system enables RNase-free sample handling, electrophoresis, staining and detection on single integrated systems. Less than 2 μl is necessary for RNA quantitation and quality check. The detection limit of this chip-based assay is approximately 100-fold lower than that of conventional agarose gel electrophoresis. Small volumes reduce the time taken to analyse and quantify RNA, the main advantage being low sample consumption, which is beneficial especially when studying fat microbiopsy samples. Microfabricated chip-based electrophoresis is therefore useful as an initial quality-control method for any RNA-related downstream application.

Applications to adipose tissue gene profiling

The low-grade inflammation that characterizes obesity and type 2 diabetes is associated with adipose tissue macrophage accumulation [3] and changes in adipokine production, thus providing a potential link between insulin resistance and endothelial dysfunction, the early stage in the atherosclerotic process [4]. An increasing number of studies have shown that weight loss is associated with an improvement in inflammatory markers together with an improvement in insulin sensitivity and endothelial function [5, 6]. Using pangenomic cDNA arrays, the transcriptome of the adipose tissue from lean and from obese women before and during a four-week Very Low Caloric Diet (VLCD) was analyzed and microarray data were cross-validated using RT-qPCR [7]. Analysis of microarray data using gene ontology [8] focused on genes with ontological criteria that shared the following syntax: immune or inflammatory response, acute phase response, cellular defense, and response to stress. Genes encoding factors that are involved in acute phase response, growth and differentiation were mostly downregulated while other categories contained equally represented up- and downregulated genes. The expression of these genes was compared to those of the adipose tissue of non-obese subjects. Cluster analysis of microarray data showed that the pattern of gene expression in obese subjects after VLCD was closer to the profile of lean subjects than to the profile of the obese. The regulation of anti-inflammatory cytokines in relation to the change in pro-inflammatory gene expression and their cellular origin was investigated using RT-qPCR. Weight loss improved the inflammatory profile of obese subjects through a decrease in pro-inflammatory factors and an increase in anti-inflammatory molecules such as interleukin 10 or IL1-receptor antagonist. Gene expression profiling was also used to compare mature adipocytes to SVF gene expression. Less than a quarter of the genes regulated during VLCD were significantly over-expressed in adipose cells, while the remaining genes were expressed either in both adipocytes and the SVF or almost exclusively in SVF. Analysis of RNA derived from microbead isolated cells from the SVF using RT-qPCR showed that a fraction of these genes were expressed in resident macrophages, which represent a large population of cells from the SVF (Fig.1). Further morphological and immunohistochemical analyses showed that "activated" macrophage cells were infiltrating adipose tissue of obese subjects [9]. A similar experimental setup led to the identification of novel biomarkers of nutritional status [10] and vascular diseases [11].

To conclude, by using human adipose tissue biopsies for gene expression profiling analysis we have described that caloric restriction-induced weight loss leads to the regulation of a wide variety of inflammation-related molecules in human subcutaneous adipose tissue. Coupled to immunohistochemistry, the isolation of the various adipose cell types has led to the characterization of the cellular source of these nutritionally regulated genes. The use of microbead-isolated cells from fresh tissue has allowed the study of purified cellular fractions. However, the low total RNA yield from some cellular fractions makes gene expression profiling somewhat difficult. Furthermore, laser capture microdissection from histological sections would now be the method of choice to overcome the complexity of the tissue. Nevertheless, attempts to extract total RNA from laser capture microdissected preparations has failed up to now. Automation would help to increase throughput and further shorten preparation time, but the need for additional preparation steps and modifications compared to conventional RNA preparation systems have, up to now, impaired the use of robotic purification systems. Finally, the combination of appropriate RNA isolation technologies, an adequate management of the microarray experiments, and proper execution of RT-quantitative PCR to crosscheck the results guarantees accurate insights in adipose tissue function.

Acknowledgements
I gratefully thank Carine Valle for valuable contributions to the work in this field. I would also like to thank Dominique Langin who has provided useful discussion and instruction.

References

1 Copland JA, Davies PJ, Shipley GL, Wood CG, Luxon BA, Urban RJ (2003) The use of DNA microarrays to assess clinical samples: the transition from bedside to bench to bedside. *Recent Prog Horm Res* 58: 25–53
2 Ruan H, Zarnowski MJ, Cushman SW, Lodish HF (2003) Standard isolation of primary adipose cells from mouse epididymal fat pads induces inflammatory mediators and down-regulates adipocyte genes. *J Biol Chem* 278: 47585–47593
3 Weisberg SP, McCann D, Desai M, Rosenbaum M, Leibel RL, Ferrante AW, Jr (2003) Obesity is associated with macrophage accumulation in adipose tissue. *J Clin Invest* 112: 1796–808
4 Xu H, Barnes GT, Yang Q, Tan G, Yang D, Chou CJ, Sole J, Nichols A, Ross JS, Tartaglia LA et al (2003) Chronic inflammation in fat plays a crucial role in the development of obesity-related insulin resistance. *J Clin Invest* 112: 1821–30
5 Cottam DR, Mattar SG, Barinas-Mitchell E, Eid G, Kuller L, Kelley DE, Schauer PR (2004) The chronic inflammatory hypothesis for the morbidity associated with morbid obesity: implications and effects of weight loss. *Obes Surg* 14: 589–600
6 Ziccardi P, Nappo F, Giugliano G, Esposito K, Marfella R, Cioffi M, D'Andrea F, Molinari AM, Giugliano D (2002) Reduction of inflammatory cytokine concentrations and improvement of endothelial functions in obese women after weight loss over one year. *Circulation* 105: 804–9
7 Clement K, Viguerie N, Poitou C, Carette C, Pelloux V, Curat CA, Sicard A, Rome S, Benis A, Zucker JD et al (2004) Weight loss regulates inflammation-related genes in white adipose tissue of obese subjects. *Faseb J* 18: 1657–69
8 Al-Shahrour F, Diaz-Uriarte R, Dopazo J (2004) FatiGO: a web tool for finding significant associations of Gene Ontology terms with groups of genes. *Bioinformatics* 20: 578–80
9 Cancello R, Henegar C, Viguerie N, Taleb S, Poitou C, Rouault C, Coupaye M, Pelloux V, Hugol D, Bouillot JL et al (2005) Reduction of macrophage infiltration and chemoattractant gene expression changes in white adipose tissue of morbidly obese subjects after surgery-induced weight loss. *Diabetes* 54: 2277–86
10 Poitou C, Viguerie N, Cancello R, De Matteis R, Cinti S, Stich V, Coussieu C, Gauthier E, Courtine M, Zucker JD et al (2005) Serum amyloid A: production by human white adipocyte and regulation by obesity and nutrition. *Diabetologia* 48: 519–28
11 Taleb S, Lacasa D, Bastard JP, Poitou C, Cancello R, Pelloux V, Viguerie N, Benis A, Zucker JD, Bouillot JL et al (2005) Cathepsin S, a novel biomarker of adiposity: relevance to atherogenesis. *Faseb J* 19: 1540–2

Sample preparation

Samples isolated by flow cytometry

Victor Appay and Martin Larsen

Immunologie Cellulaire et Tissulaire, INSERM U543, Faculté de Médecine, Hôpital Pitié-Salpêtrière, 91 Bd de l'Hôpital, 75634 Paris Cedex 13, France

Abstract

In recent years, important technological developments have provided new means to study fundamental aspects in biology and pathology. The emergence of DNA microarray technology enables us to analyze the expression levels of thousands of genes within cells, and to explore the underlying genetic causes of many human diseases. Moreover, the development of global PCR techniques now enables 1 to 10 million fold mRNA amplifications, and therefore microarray analysis from limited biological materials. Fluorescence Activated Cell Sorting (FACS) offers us the opportunity to separate cell populations according to their phenotypic or physical attributes, including rare cell types from heterogeneous mixtures, and to purify these for further characterization. In this chapter we discuss the possibility of combining these major technological advances in order to perform gene expression profiling on isolated cells of interest. The synergy of FACS and microarray technologies is bound to further our capacity to unveil the secrets of life.

Introduction

Recent years have seen striking technological advances in scientific research which have already had a considerable impact on our knowledge of biology and are expected to fundamentally transform many common medical treatments. The foundation of all biological processes is located in the genes, whose expression confers unique properties to each cell type. Studying the profile of gene expression in cells can provide unique insights into cell status, function, and behavior. Together with the result of the Human Genome Project, the emergence of DNA microarray technology has allowed scientists to analyze the expression levels of thousands of genes within cells quickly and efficiently in a single experiment; it has provided the means to understand fundamental aspects of growth and development as well as to explore the underlying genetic causes of many human diseases. Microarrays are used to assay gene expression within a single sample or to compare gene expression in two different cell types or tissue samples, such as in healthy and diseased tissue, in order to pinpoint particular genes of interest. Through the simultaneous examination of a large number of genes, the use of microarrays can help the identification of genes involved in the development of various diseases.

Microarrays in Inflammation, edited by Andreas Bosio and Bernhard Gerstmayer

However, microarray analysis are now used to ask increasingly complex questions and too many microarray studies end up in simple surveys of gene expression profiles from biological samples, with the difficulty of generating clear conclusions. A major reason for this is that samples used for microarray studies are often constituted of a mixture of cell types, from which grasping practical information can be problematic. If one wants to use this technology to study the integration of gene expression and function at the cellular level, more intricate experiments need to be performed. Primarily, this implies performing microarray analysis on samples consisting of well identified and purified cells. The separation of cells of interest from a heterogeneous mixture after their characteristics have been measured and classified has recently become possible through the development of another major technology: Fluorescence Activated Cell Sorting (FACS). Cell sorters separate a complex cell mixture into fractions of phenotypically identical cells that can be studied in isolation, and are particularly suited to collect rare cell types among a large number of non-relevant cells.

Fluorescence Activated Cell Sorting

History of FACS

In the mid 1960s, a particle separator that could fractionate particles according to their size had been developed at Los Alamos National Laboratory. At that time, Dr. Leonard Arthur Herzenberg from Stanford University, who had been performing research on lymphocytes since the late 1950s, was seeking for effective means to sort cells. Dr. Herzenberg came into the spotlight when he combined the particle separating instrument with a fluorescence detector and fluorescent-labeled monoclonal antibodies. In 1969 Dr. Herzenberg became the first in the world to successfully sort fluorescent-labeled cells that were still functional after sorting [1]. After a series of modifications in the early 1970s, he succeeded in developing a mass-producible flow cytometer called the Fluorescence Activated Cell Sorter (FACS) that performed reliable and effective live cell sorting. Dr. Herzenberg made pioneering contributions towards the development of FACS and he is considered the father of the technology, which represents a turning point in the progress of scientific research.

Principles of flow cytometry

Flow cytometry is the measurement (meter) of characteristics of single cells (cyto) suspended in a flowing saline stream. Flow cytometers accurately measure forward-scattered light (cell size), side-scattered light (cell‘s internal structure) and fluorescence intensity (cell properties) of individual fluorescent-labeled cells, which represent information on which each viable cell can be sorted [2, 3].

In a flow cytometer, cells are carried by a thin stream of saline solution emanating from a tapered nozzle. Shortly after the cell leaves the nozzle, it passes through a focused beam of laser light that illuminates each moving cell and excites fluorescent molecules on the cells. Light is scattered in all directions, but a series of temporal, spatial and chromatic filters direct and separate this light and the signals from different fluorophores to a set of detectors or photomultipliers, which convert the various intensities of the light pulses into a format suitable for computer analysis and interpretation (see Fig. 1). The total amount of forward scattered light detected is closely correlated with the cross-sectional area of the cell (i.e. its size) as seen by the laser, whereas the amount of side scattered light can indicate nuclear shape or cellular granularity. Further properties of the cell, such as surface molecules or intracellular constituents, can also be accurately quantified if the cellular marker of interest can be labeled with a fluorescent dye; for example, an antibody-fluorescent dye conjugate may be used to attach to specific surface or intracellular receptors. Other dyes have been developed which bind to particular structures (e.g. DNA, mitochondria) or are sensitive to the local chemistry (e.g. Ca^{2+} concentration, pH, etc.). The quantity of fluorescent light emitted can be correlated with the expression of the cellular marker in question.

Cell sorting

Based on all of the measured properties of fluorescence and light scatter, cells can be selected for sorting (see Fig. 1). In order to do so, a high frequency vibration is coupled to the nozzle's tip, leaving a trail of cyclical imprints onto the liquid's surface, eventually forcing the stream to separate into regularly spaced droplets. The cells soon find themselves distributed among a string of discrete droplets of saline solution. The laser beam illuminates each cell just after it emerges from the nozzle, so data is collected at this point and a decision can be made as to which cells are to be collected. If a cell meets one or more criteria set by the operator, an electrical charge may be applied to the droplets containing cells of interest to separate them from the main stream. In this manner, droplets with different cell types are directed towards separate collection vessels by a static electrical field (see Fig. 1). The cells collected are essentially unharmed by the process, so that they can be studied subsequently in *in vitro* or *in vivo* experiments, and their RNA or DNA can be readily isolated and analyzed.

FACS capacity

The instruments available for FACS have evolved considerably since the early days. Modern electronics and detectors can analyze cells at speeds greater than

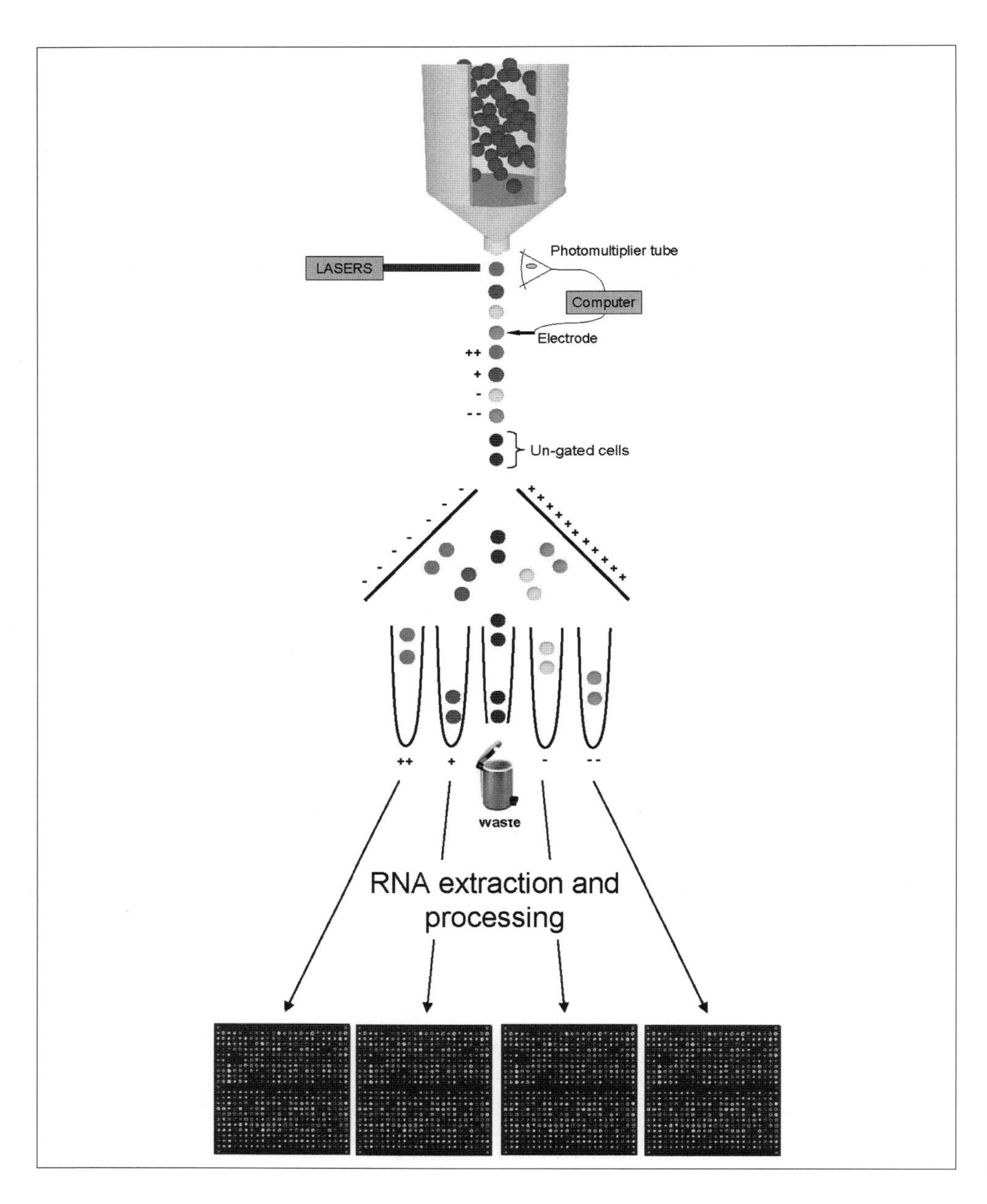

Figure 1
Schematic representation of Fluorescence Activated Cell Sorting.
Colored dots represent cells with different fluorescent signals, reflecting distinct properties (e.g. cell surface phenotype) identified though the labeling of fluorochrome conjugated monoclonal antibodies.

200,000 events per second. However the physics of drop formation and the statistics of distributing the cells among the droplets limit sort rates to about 50,000 cells per second. Final cell purity frequently reaches up to 98–99%. The combination of speed and reliable separation allows scientists to seek out very rare populations and isolate them for further study. Up to four distinct cell populations from a single cell mixture can be simultaneously sorted into tubes. Similarly, individual cells can be deposited directly into microtitre plates. Another major improvement of the technology lies in the possibility of measuring and therefore sorting cells according to a multitude of parameters at once. Each flow cytometer is usually able to detect many different fluorochromes, depending on its configuration. Nowadays, up to 15 fluorochromes may be analyzed simultaneously [4], in particular with the development of new fluorochromes, like quantum dots [5]. To achieve this, up to four separate laser beams of different colors are used so that a wide range of fluorochromes can be excited simultaneously. It has even become possible to acquire bench sorters with such capacities. These instruments can be placed in biological safety cabinets or category III laboratories in order to perform live cell sorting from infectious materials.

Impact on microarray analysis

Applications of FACS

The analysis of mixtures or bulk cell preps by microarray is no more satisfactory. MACS sorting is a simple and rapid technique for sorting a large number of cells (up to 10^7 cells) based on the expression of one marker (or even several markers by consecutive sortings) (see chapter on MACS sorting). However, FACS technically surpasses it on several points: cellular purity, simultaneous sorting of distinct cell populations based on the expression of multiple markers, and sorting of cells according to marker expression level (i.e. the relative intensity of staining). In this context, FACS fulfills the need to isolate cells with increasingly detailed phenotype using a variety of markers. FACS is used for a vast range of applications in life sciences, from basic biological disciplines such as immunology, embryology, and neuroscience, to applied biological disciplines such as medical sciences, including stem cell biology. FACS has even been applied to separate specific chromosomes prepared from cells, followed by the construction of a DNA library from each chromosome [6]. In all of these disciplines, microarray analysis can therefore be used in synergy with FACS to perform gene expression profiling on isolated cells. As long as cells can be labeled with a special fluorescence and resuspended in solution, they can be sorted, and microarray experiments can be performed. We provide examples of applications where microarray technology can be used beneficially in synergy with flow cytometry.

Gene expression profiling of populations of interest

Antigen specific T cells are one of the major components of cell mediated immunity and play a key role in the elimination of virus-infected, tumour and allograft cells. Their detailed study is necessary for our understanding of immunity efficacy and the development of satisfactory therapies (e.g. vaccines). Cells specific for one antigen or peptide are usually small populations and can be detected by various techniques. One such technique is by flow cytometry using soluble fluorescently labeled peptide/MHC complexes (usually known as tetramers), which can directly tag T cells according to their antigen specificity. Tetramer complexes are formed by four molecules of peptide-MHC class I refolded *in vitro* and bound together to a fluorochrome. Their use to stain antigen-specific $CD8^+$ T cells is based on the ability of an appropriate HLA molecule assembled with a relevant peptide (constituting the tetramer) to bind specifically to the T cell receptor of a $CD8^+$ T cell, with sufficient avidity to allow read-out by flow cytometry. The introduction of peptide-MHC tetrameric complex technology initiated a profound revolution in the field of cellular immunology [7]; it has provided a method to reliably identify specific $CD8^+$ T cells present in peripheral blood and secondary lymphoid organs [8, 9], at frequencies down to >0.01% of the total $CD8^+$ T cell population and to study their characteristics directly *ex vivo*. Such cells can readily be purified by MACS with a purity of up to 95%. However, in recent years several research groups including ours have described the very diverse differentiation states that peptide specific T cells can acquire [10, 11]. These differentiation states, which exhibit distinct functional capacity and roles, are characterized by the expression of various cell surface markers (e.g. CD27 and CD45RA) that can be identified by specific monoclonal antibodies [12]. Multiparametric FACS has therefore proven to be particularly appropriate for the study and the purification of antigen specific T cell subpopulations (see Fig. 2a). The gene profiling of the different differentiation subsets is bound to bring further enlightenment on the molecular processes involved in the development of T cells. Also, one could examine the molecular basis of antigen specific $CD8^+$ T cell subsets associated with better control of HIV replication by analyzing tetramer positive cells from donors with slow disease progression compared with cells from patients that progress rapidly towards AIDS. In the same line, one could study the particularity of antigen specific $CD8^+$ T cells infiltrated in regressing tumors *versus* circulating vaccine induced, yet ineffective, $CD8^+$ T cells that are specific for the same antigen. Identifying the causative factors that make particular cells able to control a pathogen or a tumor will provide useful keys to develop effective therapeutic strategies.

It has also become important to accurately sort cells characterized by tight intensity ranges for fluorescent markers, such as carboxyfluoroscein diacetate succinimidyl ester (CFSE), activation markers such as interleukin-2 receptor alpha (CD25) and CD40L as well as programmed death 1 (PD-1) receptor. Regulatory T cells represent a T cell subpopulation whose central role in immunity has been highlighted

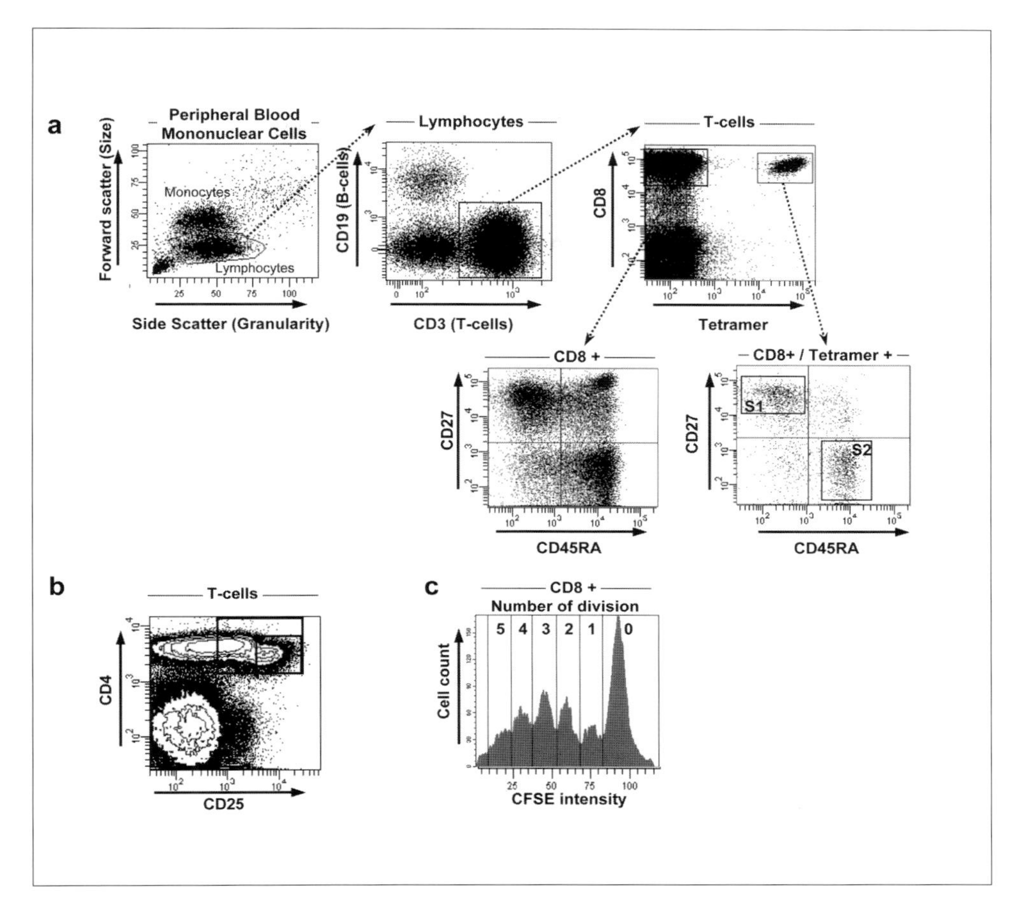

Figure 2
Examples of FACS images to illustrate gating process and selection of cell subpopulation for sorting.
a. Mononuclear cells obtained by density gradient separation from blood were labeled with fluorochrome conjugated monoclonal antibodies and placed into a flow cytometer. Side Scatter and Forward Scatter enables us to distinguish lymphocytes from monocytes. Within the lymphocytes, expression of CD19 or CD3 enables the distinction between B cells and T cells, respectively. Within the $CD3^+$ cells, antigen specific $CD8^+$ T cells can be identified using tetramers and further divided into functionally distinct subsets according to their phenotype. Antigen specific $CD8^+$ T cells present in gates S1 or S2 can therefore be sorted in order to compare their gene expression profile by microarrays.
b. Flow cytometer enables the distinction between $CD25^+$ and $CD25^{high}$ (i.e. the true regulatory T cells), and therefore their sorting for gene expression profiling.
c. Gene expression profiling from non-dividing versus proliferating cells can be performed as cells can be sorted based on the fluorescent signal provided by the CFSE dye, which accurately distinguish cells having undergone different numbers of divisions.

in recent years. The IL-2 receptor chain alpha (CD25), was historically a marker of activation, but recently it has been further recognized as a marker of regulatory T cells, on which CD25 is particularly highly expressed. However, the expression is gradual and higher expression correlates with better suppressive function of the regulatory T cells [13]. Therefore, to perform precise gene expression profiling of these cells it is necessary to discriminate accurately between $CD25^{+}$ and $CD25^{bright}$ cells (see Fig. 2b). FACS is presently the sole technique capable of distinguishing between such subsets of cells, and we will soon see studies utilizing this to generate gene profiles that can explain why $CD25^{bright}$ T regulatory cells are biologically distinct from activated $CD25^{+}$ T cells. FACS sorting followed by microarray analysis can also enable the comparison of gene expression profiles of cells after varying numbers of divisions. CFSE (succinimidyl ester of carboxyfluorsecein diacetate) is a non-toxic bright dye that is distributed equally in the cytoplasm of the cell upon simple incubation, and divides equally upon cell division. Cells undergoing multiple cell divisions will therefore show a gradual loss of CFSE fluorescence and can be sorted accordingly (see Fig. 2c). Recently, human T cells readily expanded upon stimulation with phytohaemagglutinin (PHA) were studied in a comparative analysis of gene expression profiles from FACS sorted cells having undergone no division, one, or up to four divisions. The study revealed that cells initially upregulate genes important for survival and proliferation, such as transcription factors, whereas later divisions show the upregulation of the translational machinery crucial for cell function and cytokine signaling. These data suggest that stimulated T cells need to undergo several divisions before they acquire effector functions [14, 15].

In recent years, it has also become possible to generate genetic mouse models incorporating the Enhanced Green Fluorescenc Protein (EGFP) under the regulation of promoters that otherwise regulate transcription of genes of key interest [16, 17]. These models are particularly important for the study of the influence of genes in developmental biology. Indeed, FACS isolation and microarray analysis of EGFP positive cells allow us to identify genes that are biologically linked to the gene of interest. The technique is independent of the availability of antibodies specific for the gene of interest as well as the cellular localization of the protein encoded by the gene. This technology radically improves our ability to follow gene expression in the same cell over time, since no external manipulation is needed, such as fixation or antibody staining. However, it only describes the onset of gene transcription, since the half life of EGFP and the natural gene product are rarely identical.

Microarray analysis of rare populations

The possibility of isolating well identified cells of interest in order to perform gene expression profiling nonetheless faces a particular issue, which concerns the amount of materials necessary for the analysis. Indeed, microarray experiments require a

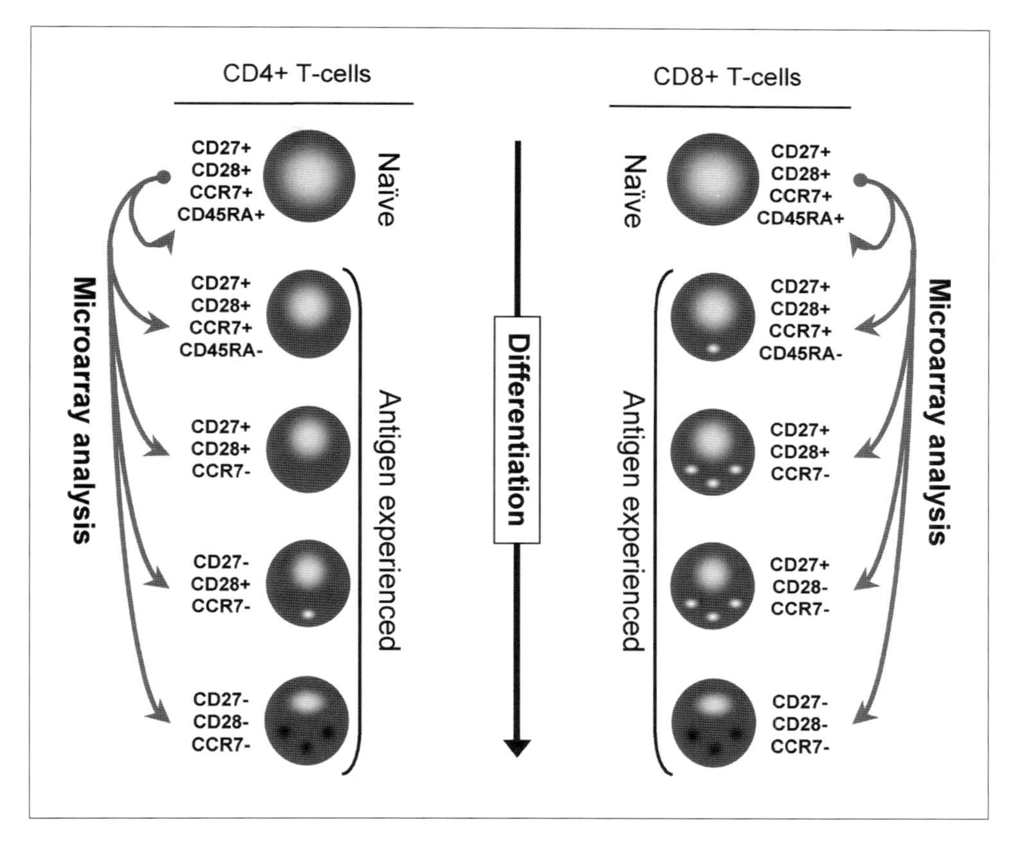

Figure 3
Microarray analysis of CD8⁺ or CD4⁺ T cell subpopulations defining distinct stages of differentiation.
Distinct CD8⁺ or CD4⁺ T cell subsets placed along a linear pathway of differentiation were identified and FACS sorted according to the expression of the cell surface receptors CD45RA, CCR7, CD28 and CD27. DNA probes from each CD8⁺ or CD4⁺ subset (1000 cells) were hybridized against a standard pool of naïve CD8⁺ or CD4⁺ respectively, as illustrated with the red arrows. Differences in gene expression profiles between naïve and antigen experienced cell subsets were obtained by microarray analysis.

substantial quantity of mRNA extracts, and therefore a large number of cells. For instance, direct reverse transcription and fluorescent labeling of mRNA uses approximately 5–10 μg of total RNA, and previous studies of primary human lymphocytes by microarray analysis used at least 10^5 cells per sample [18–21]. This may not be compatible with the isolation of rare populations, in particular from limited biological samples. Applying microarray technology to the analysis of low cell numbers sorted by FACS entails particular challenges with the downstream processing of gene expression profiling, associated with the reduced amount of starting material. Sorting of pure cell populations directly into cell lysis buffer can already participate significantly in preventing damaging mRNA, therefore allowing the use of fewer cells. Most importantly, gene expression profiling from limited material is now becoming possible, owing to the development of global PCR techniques that enable 1- to 10-million-fold mRNA amplification [22–25]. Methodologies of cDNA amplification to be applied in this context are discussed in great detail in the following chapters.

We have recently applied such technology (i.e. microarray analysis on amplified cDNA from low cell numbers, including primary T lymphocytes) to study the CD4 and CD8 lineage differentiation [26]. This technology, which reliably assesses gene expression from a small number of cells, was particularly suited for this study, since analysis had to be performed on multiple T cell subsets obtained from a unique blood sample, which can contain relatively small numbers of cells per subsets. From single blood samples of 20 ml, 1000 cells were obtained for 10 different subpopulations defining the major stages of post-thymic $CD4^+$ or $CD8^+$ T cell differentiation in humans, i.e. from naïve to highly differentiated cells [10, 11], in order to perform comparative gene expression profiling of these distinct subsets (Fig. 3). These T cell subsets were first sorted by MACS (to enrich in $CD8^+$ or $CD4^+$ T cells) and then by five-color FACS (based on the expression of CD8/CD4, CD27, CD28, CD45RA and CCR7). For both $CD4^+$ or $CD8^+$ lineages, the probes obtained for each subset were hybridized in a competitive manner against the probes from respective naïve cell pool standards (composed of three independent FACS isolated and amplified 1000 $CD4^+$ or $CD8^+$ cell samples). This analysis therefore provides information on the differential gene expression between the subsets of antigen experienced cells and the naïve cells. Bioinformatic analysis was carried out to compare the expression profiles of the different subsets of CD4 and CD8 lineage differentiation to each other. This study revealed an unexpected converging evolution of CD4 and CD8 lineages as cells become terminally differentiated, which may correspond to a common end in T cell development [26].

Concluding remarks

Biomedical research evolves and advances not only through the compilation of knowledge but also through the development of new technologies; microarrays and FACS represent two obvious examples. Microarray technology is still thought to be in its infancy and therefore has a bright future, in particular as it can take full advantage of global cDNA amplification procedures. FACS is now considered a vital and irreplaceable tool in research laboratories as well as clinics. The combination of these two major technological advances is bound to boost even further the capacities of researchers to perform gene expression profiling.

Main cell sorter manufacturers:

- Becton Dickinson, which proposes the FACSAria™ or the FACSVantage™ SE (http://www.bdbiosciences.com for information)
- Beckman coulter, which proposes the EPICS ALTRA™ or the MoFlo™ (http://www.beckman.com for information)

References

1 Hulett, H.R, W.A. Bonner, J. Barrett, L.A. Herzenberg. 1969. Cell sorting: automated separation of mammalian cells as a function of intracellular fluorescence. *Science* 166: 747–749

2 Givan, A. 2001. *Flow Cytometry: First Principles.* Wiley-Liss, New York

3 Shapiro, H.M. 2003. *Practical Flow Cytometry*, 4th edition. Wiley-Liss, Hoboken

4 Perfetto, S.P, P.K. Chattopadhyay, M. Roederer. 2004. Seventeen-colour flow cytometry: unravelling the immune system. *Nat Rev Immunol* 4: 648–655

5 Chattopadhyay, P.K, D.A. Price, T.F. Harper, M.R. Betts, J. Yu, E. Gostick, S.P. Perfetto, P. Goepfert, R.A. Koup, S.C. De Rosa et al. 2006. Quantum dot semiconductor nanocrystals for immunophenotyping by polychromatic flow cytometry. *Nat Med* 12: 972–977

6 Davies, K.E, B.D. Young, R.G. Elles, M.E. Hill, R. Williamson. 1981. Cloning of a representative genomic library of the human × chromosome after sorting by flow cytometry. *Nature* 293: 374–376

7 Altman, J.D, P.A.H. Moss, P.J.R. Goulder, D.H. Barouch, M.G. McHeyzer-Williams, J.I. Bell, A.J. McMichael, M.M. Davis. 1996. Phenotypic analysis of antigen-specific T lymphocytes [published erratum appears in *Science* 1998 Jun 19; 280 (5371): 1821]. *Science* 274: 94–96

8 Ogg, G.S, X. Jin, S. Bonhoeffer, P.R. Dunbar, M.A. Nowak, S. Monard, J.P. Segal, Y. Cao, S.L. Rowland-Jones, V. Cerundolo et al. 1998. Quantitation of HIV-1-specific cytotoxic T lymphocytes and plasma load of viral RNA. *Science* 279: 2103–2106

9 Romero, P, P.R. Dunbar, D. Valmori, M. Pittet, G.S. Ogg, D. Rimoldi, J.L. Chen, D. Lienard, J.C. Cerottini, V. Cerundolo. 1998. *Ex vivo* staining of metastatic lymph nodes by class I major histocompatibility complex tetramers reveals high numbers of antigen-experienced tumor-specific cytolytic T lymphocytes. *J Exp Med* 188: 1641–1650

10 van Lier, R.A, I.J. ten Berge, L.E. Gamadia. 2003. Human CD8(+) T-cell differentiation in response to viruses. *Nat Rev Immunol* 3: 931–939

11 Appay, V, S.L. Rowland-Jones. 2004. Lessons from the study of T-cell differentiation in persistent human virus infection. *Semin Immunol* 16: 205–212

12 Herzenberg, L.A, and S.C. De Rosa. 2000. Monoclonal antibodies and the FACS: complementary tools for immunobiology and medicine. *Immunol Today* 21: 383–390

13 Miyara, M, Z. Amoura, C. Parizot, C. Badoual, K. Dorgham, S. Trad, M. Kambouchner, D. Valeyre, C. Chapelon-Abric, P. Debre et al. 2006. The immune paradox of sarcoidosis and regulatory T cells. *J Exp Med* 203: 359–370

14 Li, Y, K.K. Wong, S. Matsueda, C.L. Efferson, D.Z. Chang, C.G. Ioannides, N. Tsuda. 2006. Mitogen stimulation activates different signaling pathways in early- and late-divided T cells as revealed by cDNA microarray analysis. *Int J Mol Med* 18: 1127–1139

15 Tanaka, Y, H. Ohdan, T. Onoe, T. Asahara. 2004. Multiparameter flow cytometric approach for simultaneous evaluation of proliferation and cytokine-secreting activity in T cells responding to allo-stimulation. *Immunol Invest* 33: 309–324

16 Kawaguchi, A, T. Miyata, K. Sawamoto, N. Takashita, A. Murayama, W. Akamatsu, M. Ogawa, M. Okabe, Y. Tano, S.A. Goldman, H. Okano. 2001. Nestin-EGFP transgenic mice: visualization of the self-renewal and multipotency of CNS stem cells. *Mol Cell Neurosci* 17: 259–273

17 Maric, D, J.L. Barker. 2004. Neural stem cells redefined: a FACS perspective. *Mol Neurobiol* 30: 49–76

18 Willinger, T, T. Freeman, H. Hasegawa, A.J. McMichael, M.F. Callan. 2005. Molecular signatures distinguish human central memory from effector memory CD8 T cell subsets. *J Immunol* 175: 5895–5903

19 Holmes, S, M. He, T. Xu, and P.P. Lee. 2005. Memory T cells have gene expression patterns intermediate between naive and effector. *Proc Natl Acad Sci USA* 102: 5519–5523

20 Boutboul, F, D. Puthier, V. Appay, O. Pelle, H. Ait-Mohand, B. Combadiere, G. Carcelain, C. Katlama, S.L. Rowland-Jones, P. Debre et al. 2005. Modulation of interleukin-7 receptor expression characterizes differentiation of CD8 T cells specific for HIV, EBV and CMV. *Aids* 19: 1981–1986

21 Menzel, O, M. Migliaccio, D.R. Goldstein, S. Dahoun, M. Delorenzi, N. Rufer. 2006. Mechanisms regulating the proliferative potential of human $CD8^+$ T lymphocytes overexpressing telomerase. *J Immunol* 177: 3657–3668

22 Iscove, N.N, M. Barbara, M. Gu, M. Gibson, C. Modi, N. Winegarden. 2002. Representation is faithfully preserved in global cDNA amplified exponentially from sub-picogram quantities of mRNA. *Nat Biotechnol* 20: 940–943

23 Smith, L, P. Underhill, C. Pritchard, Z. Tymowska-Lalanne, S. Abdul-Hussein, H. Hilton, L. Winchester, D. Williams, T. Freeman, S. Webb, A. Greenfield. 2003. Single primer amplification (SPA) of cDNA for microarray expression analysis. *Nucleic Acids Res* 31:e9

24 Hartmann, B, F. Staedtler, N. Hartmann, J. Meingassner, H. Firat. 2006. Gene expression profiling of skin and draining lymph nodes of rats affected with cutaneous contact hypersensitivity. *Inflamm Res* 55: 322–334

25 Jensen, K.B, and F.M. Watt. 2006. Single-cell expression profiling of human epidermal stem and transit-amplifying cells: Lrig1 is a regulator of stem cell quiescence. *Proc Natl Acad Sci USA* 103:11958–11963

26 Appay, V, A. Bosio, S. Lokan, Y. Wiencek, C. Biervert, D. Kusters, D. Devevre, D.E. Speiser, P. Romero, N. Rufer, S. Leyvraz. 2007. Sensitive gene expression profiling of human T-cell subsets reveals parallel post-thymic differentiation for $CD4^+$ and $CD8^+$ lineages. *J Immunol;* in press

Sample preparation

Magnetic cell sorting

Jürgen Schmitz

Miltenyi Biotec GmbH, Friedrich-Ebert-Straße 68, 51429 Bergisch Gladbach, Germany

Abstract

Diverse cell types are known to contribute to pathology in inflammatory diseases such as rheumatoid arthritis. Analyzing inflammatory disease-related alterations in gene expression in different cell types requires cell purification without influencing the gene expression profile. Magnetic cell sorting has become a standard method to isolate almost any cell type in complex cell mixtures such as peripheral blood. The major differences between the magnetic cell separation systems that are currently used are the composition and size of the magnetic particles used for cell labeling and the mode of magnetic separation. Large (0.5–5 μm diameter) magnetic particles have several disadvantages when compared to small (20–150 nm diameter) particles [1], including slower cell-bead reaction kinetics, a higher degree of non-specific cell-bead interaction, higher risk of non-specific entrapment of cells in particle aggregates, and adverse effects of particles on viability and optical properties of labeled cells. Here, we focus on applications of magnetic cell sorting with MACS® technology [1–5] that makes use of nanometer-size super-paramagnetic particles and high gradient magnetic fields.

Principle of high gradient magnetic cell sorting using MACS® technology

The general principle of magnetic cell sorting with MACS® is outlined in Figure 1. A suspension of cells is specifically immunomagnetically labeled using small super-paramagnetic MicroBeads that are typically directly covalently conjugated to a monoclonal antibody (mAb) or a ligand that is specific for a certain cell type. After magnetic labeling, the cells are passed through a separation column that is placed in the external magnetic field of a strong permanent magnet. The magnetizable column matrix serves to create a high gradient magnetic field that acts as a specific magnetic cell filter: Magnetically labeled cells are retained on the column, while unlabeled cells flow through. The matrix is rapidly demagnetized after removal of the column from the external magnetic field and the retained cells can easily be eluted. Both cell fractions – magnetically labeled and non-magnetically labeled cells – are imme-

Microarrays in Inflammation, edited by Andreas Bosio and Bernhard Gerstmayer

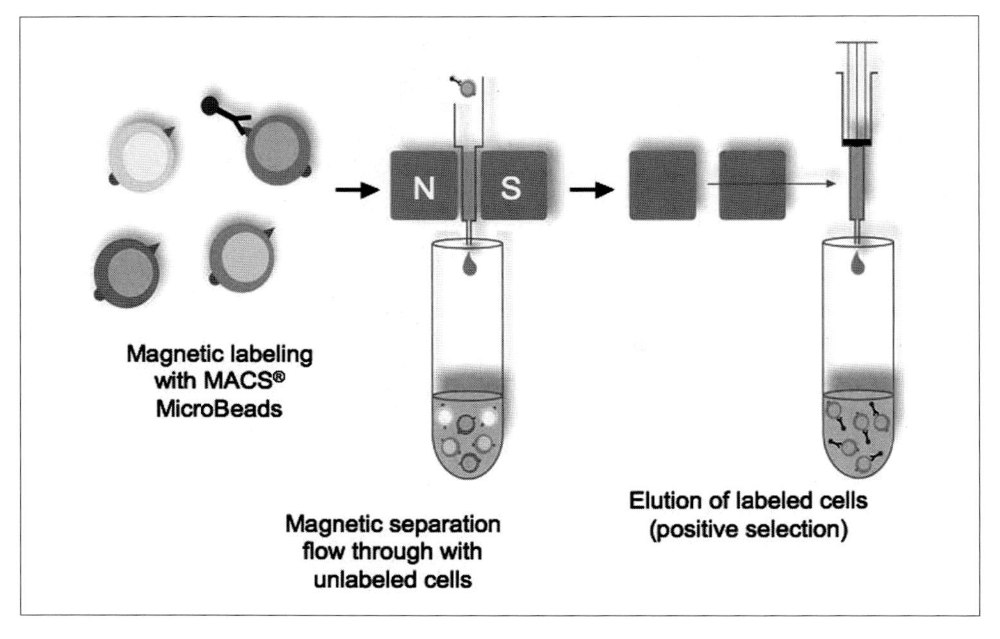

Figure 1.
Principle of high gradient magnetic cell sorting. Cells of interest are specifically magnetically labeled with small super-paramagnetic MicroBeads in a short incubation step. The mixture of magnetically labeled and non-labeled cells is passed over a separation column placed in the magnetic field of a strong permanent magnet. Magnetically labeled cells are retained on the column, while unlabeled cells flow through (negative fraction). After removal of the column from the external magnetic field the retained cells (positive fraction) can easily be eluted by rinsing the column with buffer.

diately suitable for further use, e.g. flow cytometry, molecular biology, cell culture, transfer into animals or even human cell therapy.

Super-paramagnetic MicroBeads

MACS® technology uses magnetic particles made of dextran and iron oxide with a diameter between 20 and 100 nm [1–5]. They are small enough to be colloidal, i.e. they form a stable solution due to the random molecular bombardment of Brownian motion. The magnetically susceptible core, a cluster of iron oxide microcrystals, is too small (approximately 10 nm diameter) to have a defined magnetic orientation, i.e. it is super-paramagnetic. Super-paramagnetism refers to materials that become magnetic in the presence of an external magnet, but revert to a non-magnetic state when

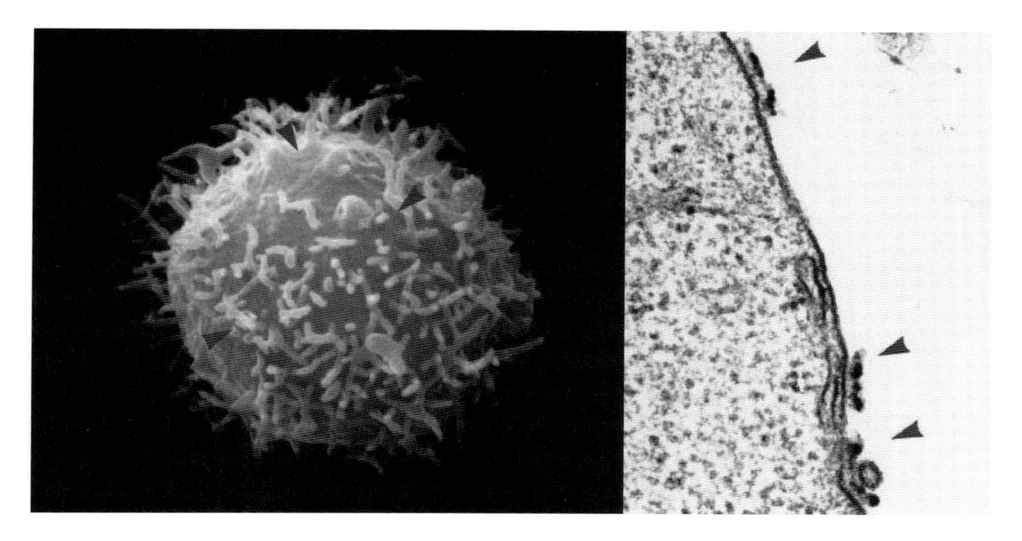

Figure 2.
Scanning (left picture) and transmission (right picture) electron micrograph of a $CD8^+$ T cell isolated by MACS® using CD8 Ab-conjugated super-paramagnetic MicroBeads (EM courtesy of Prof. Groscurth, Zürich, Switzerland). Some super-paramagnetic MicroBeads are marked with arrows. They are about 50 nm in diameter, form colloidal solutions, and are biodegradable.

the external magnet is removed. Particles with a residual magnetic moment clump together quickly. The dextran matrix of the MicroBeads is suitable for a variety of bioconjugate chemistries [1, 6]. Numerous mAb, fluorochromes, oligonucleotides, and various other molecules have all been covalently linked to MicroBeads [1, 7–9].

The use of very small iron-dextran particles is the basis for several unique features of MACS® technology. The iron-dextran composition and small surface per particle minimize unspecific binding and allow an enrichment of cells of more than 10,000-fold, even at frequencies below 10^{-8}. Particles bound to the cell do not change its optical properties, i.e. they do not change the scatter properties of cells in flow cytometric analysis and do not influence light microscopic appearance. The MicroBeads are only visible by transmission and scanning electron microscopy (Fig. 2). Cells labeled with MicroBeads have been used for numerous functional *in vitro* assays, experimental transfers into animals [10], and even for therapeutic transplants in humans [5, 11, 12]. Effects on the functional status of cells by magnetic labeling with MicroBeads are primarily dependent on the target cell surface antigen that is used and on the degree of its cross-linking by Ab or ligands conjugated to the MicroBeads, but not on the MicroBeads themselves. Iron-dextran particles are also used as contrast agents for magnetic resonance imaging. There are a variety of so-called MION (monocrystalline iron oxide nanocompounds) or SPIO (Super-

Paramagnetic Iron Oxide) reagents available on the market: Feridex®, Endorem®, GastroMARK®, Lumirem®, Sinerem®, and Resovist®. Iron-dextran particles are also injected intravenously or intramuscularly for the treatment of iron deficiency anemia (InFeD® or Dexferrum®). Iron oxide particles are non-toxic. The LD50 of iron-dextran is not less than 500 mg/kg in the mouse.

High gradient magnetic cell separation devices

The amount of magnetically susceptible material bound to cells labeled with superparamagnetic MicroBeads is very small because of the small volume of the particles. Cells with such a small magnetic moment cannot be separated in magnetic fields of conventional geometries but rather in high gradient magnetic fields. The MACS® system uses high gradient magnetic cell separation units consisting of a strong permanent magnet of 0.4-1 Tesla and a separation column with a matrix of ferromagnetic steel wool or iron spheres. When columns are placed between poles of an external permanent magnet, the homogeneous field is disturbed and high magnetic gradients are generated in the vicinity of the ferromagnetic matrix material. In their immediate neighborhood the ferromagnetic structures generate magnetic forces of more than 10^4 T/m as compared to about 10 T/m in conventional geometries [1]. Once the column is removed from the magnet, the column matrix rapidly demagnetizes and the cells that are retained can be eluted easily and completely by simply rinsing the column with buffer. To avoid corrosion and potential damage to the cells through direct contact with the matrix material, the ferromagnetic matrix is coated with a thin, biocompatible plastic polymer layer. MACS® technology allows fast processing of up to 120×10^9 cells, with as many as 40×10^9 being magnetically labeled. This is in striking contrast to fluorescence-activated cell sorting (FACS), where cells are analyzed one by one and sorted one after the other. Thus, its capacity is limited by the frequency of analysis and sorting, which is at most 5×10^4 cells per second, i.e. 120×10^9 cells in 2.4×10^6 seconds (about 28 days). Magnetic cell separation based on MACS® technology can easily be performed using the autoMACS®Pro Separator, a new benchtop automated magnetic cell sorter. Clinical-scale magnetic selection of target cells or depletion of non-target cells in a closed and sterile system based on MACS® technology can be performed using the CliniMACS® Plus Instrument.

Magnetic cell separation strategies

Two basic magnetic cell separation strategies should be considered: positive selection and depletion. The optimal strategy for any specific cell separation depends on the frequency of target cells in the cell sample, the immunophenotype of the target cell population as compared to the other cells in the sample, the availability of reagents, and a full consideration of how the target cells are to be used, including any requirements with respect to purity, yield, and activation status.

Positive selection, i.e. labeling and enrichment of target cells, requires at least one cell surface marker that is specific for the target cells. Positive selection is particularly well suited for the isolation of rare cells, such as hematopoietic stem cells, from complex cell mixtures, such as blood cells. The column just needs to be large enough to retain the target cells, allowing effective concentration of the rare cells on the column. If very rare cells are isolated (frequency of about 0.1%), nonspecific cell retention may be equal to or exceed retention of target cells, i.e. purities may be below 50%. To further enrich target cells to purities of typically more than 95%, it is possible, however, to elute the retained cells and repeat the magnetic separation using a second column.

The functional status of the cells can be influenced interdependent with the cell type, the target surface molecules used upon magnetic labeling, and the labeling moiety of the MicroBeads (mAb or ligand). This is a problem inherent to labeling with Ab or ligands that recognize and potentially crosslink cell surface receptors, which may thus induce or inhibit signal transduction. Labeling with Ab-conjugated MicroBeads has no additive effect compared to labeling with the very same non-conjugated Ab. The contrary may be the case: Labeling with Ab-conjugated MicroBeads is performed under non-saturating labeling conditions, i.e. only a minor proportion of target cell surface molecules is labeled and triggered, whereas mAb labeling is typically performed under saturating labeling conditions, i.e. all target cell surface molecules are labeled and triggered.

Depletion, i.e. labeling and removal of non-target cells to obtain the cells of interest, ensures that target cells remain unlabeled and minimizes potential effects on the functional status of cells. This strategy may sometimes be advantageous if functional studies, such as T cell activation studies, are to be performed with the target cells. Removing non-target cells by depletion is often preferred when target cells are heterogeneous and specific cell surface markers for all target cells are not known. Typical examples of preferred depletion approaches are the selective elimination of human foreskin fibroblasts (HFFs) or mouse embryonic fibroblasts (MEFs) from human embryonic stem cell cultures (removal of feeder cells), the enrichment of non-hematopoietic tumor cells from blood and bone marrow of cancer patients by leukocyte depletion, the depletion of cancer cells from autologous stem cell grafts, and the depletion of T cells from allogeneic stem cell grafts. However, even very rare and defined target cells, such as plasmacytoid dendritic cells, are nowadays routinely isolated from complex cell mixtures such as blood cells, with high purities and yields by using depletion just as well as by positive selection. Labeling of all non-target cells for depletion is often achieved with an optimized panel of sometimes more than 10 biotinylated mAb directed against all non-target cells and anti biotin mAb-conjugated MicroBeads.

In many cases, target cells cannot be specifically defined by a single cell surface marker but by multiple cell surface antigens. There are several different strategies for multi-parameter magnetic cell sorting. First, depletion may be followed by positive selection. The non-retained cells from the first separation are again magnetically labeled and enriched on a second column. In order to obtain the highest purities, different magnetic separation stringencies may be used for the two separations. The depletion step is performed on a column optimized for cell depletion and highest retention rates for labeled cells, while the enrichment step is performed on a column optimized for positive selection with lower retention rates for (non-specifically) weakly labeled cells. This reduces the probability that labeled cells will be carried over from the first separation step into the second. A typical example of a routinely used depletion followed by positive selection approach is the isolation of regulatory T helper cells from blood by depletion of all cells except for $CD4^+$ T cells, followed by positive selection of $CD25^+$ regulatory T helper cells from pre-enriched $CD4^+$ T cells [13]. Next, a positive selection step may be followed by either another positive selection or a depletion step. This can be accomplished by using colloidal super-paramagnetic particles that can be rapidly released from the cell (Multisort MicroBeads) using an enzyme (MultiSort Release Reagent). Since the specificity of the enzyme is unique to the magnetic particles, cell surface molecules are not modified. Typical candidates for double-positive selection approaches are $CD19^+$ $CD1c^+$ myeloid dendritic cells, $CD56^+$ $CD3^+$ T cells, $CD4^+$ $CD62L^+$ naïve T helper cells, etc. [14]. The concept of positive selection followed by depletion is very attractive for the depletion of contaminating tumor cells or alloreactive T cells from purified $CD34^+$ hematopoietic progenitor cells for therapeutic autologous or allogeneic stem cell grafting. Positive selection followed by either positive selection or depletion can also be achieved by using two different platforms for magnetic cell separation: Positive selection by high gradient magnetic cell separation using MicroBeads and high gradient magnetic cell separation devices followed by positive selection or a depletion using MACSiBeads (super-paramagnetic beads of 3.5 µm diameter) and a MACSiMag™ separator (a separator designed for cells labeled with MACSiBeads). MACSiBead-labeled cells but not MicroBead-labeled cells are retained using the MACSiMAG™ separator.

Preparation of cells

For good magnetic separation it is essential to have a suspension of single cells that is stable for several hours without forming clumps or aggregates. Methods for preparation of single-cell suspensions from tissue are the same as those for flow cytometry: disruption by enzymatic or non-enzymatic means. Single cell suspensions from tissues can easily be prepared by enzymatic as well as non-enzymatic means using the gentleMACS™ Dissociator, a new benchtop instrument for dissociation of tissues. Since aggregates or clumps contain mixtures of different cell types, their formation adversely affects purity and yield of cells of interest. Cells should be well re-suspended, especially after each centrifugation step. Buffer supplementation with EDTA and bovine serum albumin (BSA) or serum may aid in reducing aggregation. Removal of dead cells prior to magnetic labeling can be an important step, because dead cells release DNA, thus supporting formation of cell clumps. In addition, dead cells often become non-specifically labeled by Ab-MicroBead conjugates. Dead cells and cell debris can be removed, e.g. by Ficoll Paque® density gradient centrifugation or by magnetic depletion using dead cell removal MicroBeads [15].

Magnetic labeling strategies

For many cell surface markers on human, mouse, rat, and non-human primate cells, one-step reagents, where a cell surface-antigen specific mAb is directly conjugated to the MicroBeads, are available. Two-step magnetic labeling is the strategy of choice when MicroBeads for one-step labeling are not available, when a panel of Ab directed against multiple cell surface antigens is used, or when two-step magnetic labeling is significantly more efficient as compared to one-step labeling. Three different indirect magnetic labeling systems are often used: (a) primary Ab and anti immunoglobulin Ab-conjugated MicroBeads, (b) biotinylated primary Ab and streptavidin-conjugated MicroBeads or anti biotin mAb-conjugated MicroBeads and (c) flourochrome-conjugated (e.g. FITC-conjugated) primary Ab and anti fluorochrome Ab-conjugated MicroBeads (e.g. anti FITC Ab MicroBeads).

Ab-conjugated MicroBeads are typically used at a dilution of 1:5 or 1:10, and cells are typically incubated with MicroBeads at a cell concentration of 10^8 per ml for about 15 minutes at 2–8°C. Incubations on ice require increased incubation times for efficient magnetic labeling. Higher temperatures and longer incubation times may decrease the specificity of the Ab-MicroBead labeling. To avoid capping of the Ab-conjugated MicroBeads from the cell surface, one should work fast and keep the cells cold. An Fc receptor blocking reagent, e.g. human serum IgG for blocking of human Fc receptors, may be added to the cells directly before magnetic labeling to prevent Fc receptor-mediated labeling of non-target cells.

Two-step magnetic labeling should be optimized by careful titration of the primary Ab to avoid background staining of non-target cells. If a panel of Ab directed against multiple cell surface antigens is used, each Ab must be titrated separately. Typically, cell surface antigens are labeled to a high degree of saturation with most primary Ab within 5 to 10 minutes. Longer incubation times may decrease the specificity of Ab labeling.

MACS® control: Controlling the efficiency of magnetic cell separation

Magnetically labeled cells can be stained with fluorochrome-conjugated Ab and analyzed by flow cytometry. The small super-paramagnetic MicroBeads do not interfere with the light scattering or immunofluorescence of cells. The MACSQuant® analyzer, a new digital benchtop flow cytometer, supports MACS® control applications by special analysis features.

In general, using one-step reagents, only a minor proportion of the target antigens on the cell surface are labeled with MicroBeads under normal labeling conditions, i.e. even fluorochrome conjugated Ab with the same specificity (same Ab clone) can be used for fluorescent staining. However, if the same or similar specificities are used, fluorescent labels should be introduced after magnetic labeling to avoid inhibition of magnetic labeling.

In most cases the primary Ab can be conjugated to a fluorochrome when a two-step approach for magnetic labeling is used. However, if the primary Ab is coupled to a bulky fluorochrome such as phycoerythrin, binding of secondary anti immunoglobulin Ab-conjugated MicroBeads may be impaired because of steric hindrance. FITC-conjugated, PE-conjugated, and APC-conjugated primary Ab can also be used in combination with anti FITC Ab-, anti PE Ab-, and anti APC-conjugated MicroBeads, respectively - an increasingly popular and reliable two-step approach to label cells both magnetically and fluorescently. When employing biotinylated primary Ab, fluorescent labeling can be performed by adding an avidin-fluorochrome-conjugate after magnetic labeling with streptavidin-conjugated or anti biotin mAb-conjugated MicroBeads.

The performance of a magnetic cell separation is typically described by the purity, yield, and viability of the isolated target cells. However, since the purity of the isolated target cells is strongly dependent on the frequency of target cells in the original cell sample, determining the enrichment and depletion rates for labeled cells is extremely helpful to optimize a magnetic separation.

The enrichment rate represents the average number of negative cells passing through the column per negative cell retained non-specifically (assuming that all positive cells are retained).

$$\text{Enrichment rate } (f_E) = \frac{\%\text{ neg. cells in orig. sample}}{\%\text{ pos. cells in orig. sample}} \times \frac{\%\text{ pos. cells in pos. fraction}}{\%\text{ neg. cells in pos. fraction}}$$

This value will be low if cells are non-specifically labeled and retained. Magnetic labeling and cell separation based on MACS® technology allows the achievement of enrichment rates of more than 20,000. This means that cells with a frequency of 0.1% can easily be enriched to 95% or more, cells with a frequency of 1% can easily be enriched to 99.5% or more, and cells with a frequency of 10% can easily be enriched to 99.96% or more.

The depletion rate represents the average number of positive cells being retained per positive cell passing through the column (assuming that all negative cells pass through).

$$\text{Depletion rate } (f_D) = \frac{\text{\% pos. cells in orig. sample}}{\text{\% neg. cells in orig. sample}} \times \frac{\text{\% neg. cells in pos. fraction}}{\text{\% pos. cells in pos. fraction}}$$

This value will be low if magnetic labeling is not sufficiently strong. Magnetic labeling and cell separation based on MACS® technology allows the achievement of depletion rates of more than 1,000,000, i.e. >4 log(10) removal of positive cells. This means that cells with a frequency of 10.0% can easily be depleted to 0.001% or less, cells with a frequency of 50% can easily be depleted to 0.01% or less, and cells with a frequency of 90% can easily be depleted to 0.09% or less.

Conclusion

Gene expression profiles of mixtures of cells originate from the gene expression profiles of each cell type and their relative abundance in the overall cell sample being analyzed. Rare cell population contribute barely to the overall gene expression profile of cell mixtures, except for genes which are highly expressed in the rare cell population and not at all or only weakly expressed in all other cells. The lower the frequency of a cell population is the higher is the risk that the gene expression profile is considerably masked by the expression profile of the other cells. Several methodologies can be employed to enrich the target cells for microarray analyses. These include magnetic sorting and laser capture microdissection. Magnetic cell sorting with the MACS® technology as described here allows to obtain highly pure cell populations without influencing their gene expression profile.

References

1 Miltenyi S, Müller W, Weichel W, Radbruch A (1990) High gradient magnetic cell separation with MACS. *Cytometry* 11: 231–238

2 Radbruch A, Mechtold B, Thiel A, Miltenyi S, Pflüger E (1994) High gradient magnetic cell sorting. *Methods in Cell Biology* 42:387–403

3 Kantor AB, Gibbons I, Miltenyi S, Schmitz J (1997) Magnetic cell sorting with colloidal super-paramagnetic particles. In: D Recktenwald, A Radbruch (eds): *Cell separation methods and applications*. Marcel Dekker Inc, New York Basel Hong Kong, 153–173

4 Schmitz J, Miltenyi S (1999) High gradient magnetic cell sorting. In: A Radbruch (ed): *Flow cytometry and cell sorting*. Springer-Verlag, Berlin, 218–247

5 Apel M, Heinlein UAO, Miltenyi S, Schmitz J, Campbell JDM (2007) Magnetic cell separation for research and clinical applications. In: W Andrä, H Nowak (eds): *Magnetism in medicine*. Wiley-VHC, Weinheim, 571–595

6 Molday RS, MacKenzie D (1982) Immunospecific ferromagnetic iron-dextran reagents for the labeling and magnetic separation of cells. *J Immunol* Methods 52: 353–367

7 Kato K, Radbruch A (1993) Isolation and characterization of CD34^{+} hematopoietic stem cells from human peripheral blood by high gradient magnetic cell sorting. *Cytometry* 14:384–392

8 Brosterhus H, Brings S, Leyendeckers H, Manz RA, Miltenyi S, Radbruch A, Assenmacher M, Schmitz J (1999) Enrichment and detection of live antigen-specific CD4^{+} and CD8^{+} T cells based on cytokine secretion. *Eur J Immunol* 29: 4053–4059

9 Leyendeckers H, Odendahl M, Löhndorf A, Irsch J, Spangfort M, Miltenyi S, Hunzelmann N, Assenmacher M, Radbruch A, Schmitz J (1999) Correlation analysis between frequencies of circulating antigen-specific IgG-bearing memory B cells and serum titers of antigen-specific IgG. *Eur J Immunol* 29: 1406–1417

10 Schmitz J, Thiel A, Kühn R, Rajewsky K, Müller W, Assenmacher A, Radbruch A (1994) Induction of interleukin 4 (IL-4) expression in T helper (Th) cells is not dependent on IL-4 from non-Th cells. *J Exp Med* 179: 1349–1353

11 Handgretinger R, Lang P, Schumm M, Taylor G, Neu S, Koscielnak E, Niethammer D, Klingebiel T (1998) Isolation and transplantation of autologous peripheral CD34^{+} progenitor cells highly purified by magnetic-activated cell sorting. *Bone Marrow Transplant* 21: 987–993

12 Lang P, Schumm M, Taylor G, Klingebiel T, Neu S, Geiselhart A, Kuci S, Niethammer D, Handgretinger R (1999) Clinical scale isolation of highly purified peripheral CD34^{+} progenitors for autologous and allogeneic transplantation in children. *Bone Marrow Transplant* 24: 583–9

13 Fallarino F, Grohmann U, Hwang KW, Orabona C, Vacca C, Bianchi R, Belladonna ml, Fioretti MC, Alegre ml, Puccetti P (2003) Modulation of tryptophan catabolism by regulatory T cells. *Nat Immunol* 4:1206–1212

14 Santourlidis S, Trompeter HI, Weinhold S, Eisermann B, Meyer KL, Wernet P, Uhrberg M (2002) Crucial role of DNA methylation in determination of clonally distributed killer cell Ig-like receptor expression patterns in NK cells. *J Immunol* 169:4253–4261

15 Kinoshita N, Hiroi T, Ohta N, Fukuyama S, Park EJ, Kiyono H (2002) Autocrine IL-15 mediates intestinal epithelial cell death *via* the activation of neighboring intraepithelial NK cells. *J Immunol* 169:6187–6192

Single and rare cell analysis – amplification methods

T7 based amplification protocols

Stephen D. Ginsberg

Center for Dementia Research, Nathan Kline Institute, and Departments of Psychiatry and Physiology & Neuroscience, New York University School of Medicine, Orangeburg, NY 10962, USA

Abstract

RNA amplification is a series of molecular manipulations designed to amplify genetic signals from minute quantities of starting input RNA acquired *via* microaspiration, such as laser capture microdissection (LCM) for downstream genetic analyses that include microarrays and real-time quantitative PCR (qPCR). Biological samples harvested for RNA amplification procedures can originate from a myriad of *in vivo* and *in vitro* tissue sources. Moreover, a variety of fresh, frozen, and fixed tissues can be employed, and these tissues can be processed for histochemistry and/or immunocytochemistry prior to microdissection for RNA amplification, allowing for tremendous cell type and tissue specificity. Essentially, RNA amplification produces quantities of RNA through *in vitro* transcription (IVT). The present chapter will illustrate several RNA amplification schemes including amplified antisense RNA (aRNA) amplification and terminal continuation (TC) RNA amplification that employ bacteriophage transcription promoters such as T7 to enable IVT and subsequent generation of amplified RNA products for downstream genetic applications.

Introduction

Conventional techniques utilized in molecular and cellular biology enable research scientists to evaluate gene expression levels across a plethora of normative situations, experimental paradigms, and pathological conditions. These relatively low throughput methods typically provide presence detection and quantitation of individual transcripts one at a time (or a few at a time) including: Southern analysis (DNA detection), Northern analysis (RNA detection), polymerase-chain reaction (PCR; DNA detection), reverse-transcriptase-PCR (RT-PCR; RNA detection), ribonuclease (RNase) protection assay, and *in situ* hybridization, among others. Advances in high-throughput genomic based technologies have allowed the assessment of dozens to hundreds to thousands of genes simultaneously in a coordinated fashion. The potential to understand mechanisms underlying physiological processes and disease pathogenesis has expanded geometrically with the advent and utility of these technologies.

Microarrays in Inflammation, edited by Andreas Bosio and Bernhard Gerstmayer

Single cell gene expression as well as gene expression analysis of homogeneous populations of cells (termed population cell analysis) is an arduous task that demands a multidisciplinary approach including molecular biology, cell biology, anatomy, and biomedical engineering. Individual cell types are likely to have unique patterns, or a mosaic of gene and protein expression under normative conditions that is likely to be altered in pathological states. Indeed, the molecular basis of why distinct populations of cells are vulnerable to degeneration, often termed "selective vulnerability", can be elucidated by discrete cell analysis more readily than by utilizing regional and total tissue preparations [1–3]. A detailed pattern of genomic and proteomic expression in a subpopulation of homogeneous cells or single cells is more likely to be informative than a similar pattern in a whole tissue admixture, assuming the target population is well-defined. With the application and increased utility of modern molecular and cellular techniques, it is now possible to isolate and study RNA species, (as well as genomic DNA and proteins) from microdissected tissue sources. Unfortunately, the quantity of RNA harvested from a single cell, estimated to be approximately 0.1–1.0 picograms, is not sufficient for standard RNA extraction procedures [4, 5].

Microaspiration

Microdissection encompasses a useful set of techniques that is performed to enable downstream gene expression profiling. Provided that procedures are utilized on fresh, frozen, or well-fixed tissue sections and RNase free conditions are employed, both immunocytochemical and histochemical methods can be used to identify specific cell(s) of interest [6–8]. Discrimination and isolation of adjacent cell types from one another is critical because this enables the selection of relatively pure populations of individual cells and/or populations for subsequent analysis while avoiding potential contamination. Specifically, laser capture microdissection (LCM) is a strategy for acquiring histochemically and/or immunocytochemically labeled cells from *in vivo* and *in vitro* sources [9–12]. LCM employs a laser source directly on the cell(s) of interest for the purpose of microaspiration [13, 14]. A non-destructive, low-power, near-infrared laser pulse is directed through a microfuge cap at the target cell (a positive extraction strategy). The pulsed laser energy causes localized activation of a specialized thermoplastic film that adheres to the target cell. Raising the thermoplastic cap separates targeted cells, now attached to the film, from surrounding undisturbed tissue. Populations of cells attached to the cap are suitable for microscopic examination and downstream genetic analysis.

Rationale for RNA amplification

An RNA amplification technique is often required when attempting expression profiling from single neurons, groups of neurons, or microdissected regions. PCR-based amplification methods are not optimal, as exponential amplification can skew the original quantitative relationships between genes from an initial population [15]. Linear RNA amplification-based procedures increase RNA by directly amplifying the initial template sequence. Resultant amplified products are representative of the original mRNA expression levels. RNA amplification is achieved through *in vitro* transcription (IVT). In addition to the linear amplification of minute amounts of input RNA, RNA amplification procedures can increase the sensitivity of gene expression profiling on microarrays even though a relatively large quantity of starting material is available [16, 17]. In the presence of a proper buffering system, nucleotide triphosphates (NTPs) and a bacteriophage transcription promoter sequence (e.g., T7, T3, or SP6) the respective RNA polymerase will synthesize hundreds to thousands of high fidelity copies of the cDNA template [18, 19]. There are numerous RNA amplification schemes available, each with distinct advantages and potential limitations [20, 21].

RNA amplification strategies: aRNA

A well known linear amplification procedure, termed amplified antisense RNA (aRNA) amplification, developed by Eberwine and colleagues [15, 22, 23], consists of a T7 RNA polymerase based amplification procedure (Fig. 1). aRNA amplification entails hybridization of an oligonucleotide primer consisting of an oligo d(T) sequence and a T7 RNA polymerase promoter sequence {oligo d(T)T7)} to mRNAs and subsequent generation of an mRNA-cDNA hybrid by reverse transcriptase [24, 25]. A functional T7 RNA polymerase promoter is formed upon conversion of the mRNA-cDNA hybrid to double stranded cDNA. aRNA synthesis occurs with the addition of T7 RNA polymerase and NTPs. Each round of aRNA results in an approximate thousand-fold amplification from the original amount of each polyadenylated {poly(A)+} mRNA in the sample [22, 23]. Two rounds of aRNA are typically employed for desired downstream analyses, including microarrays. aRNA products tend not to be of full length, and are biased towards the 3' end of the transcript because of the initial priming at the poly(A)+ RNA site [4, 15]. This 3' bias exists for all amplified aRNA products and relative levels of gene expression can be compared [4, 26, 27]. Our laboratory has generated successful studies on microaspirated animal model and human tissues utilizing aRNA in combination with a wide variety of array platforms [2, 28–31]. Modifications of the original aRNA procedure have been performed [32–34], and a variety of strategies have been developed by independent laboratories to improve aRNA amplification efficiency [33, 35–38]. Several kits that use aRNA technology to amplify small amounts of RNA are commercially available, although single cell analysis is not always possible with these systems.

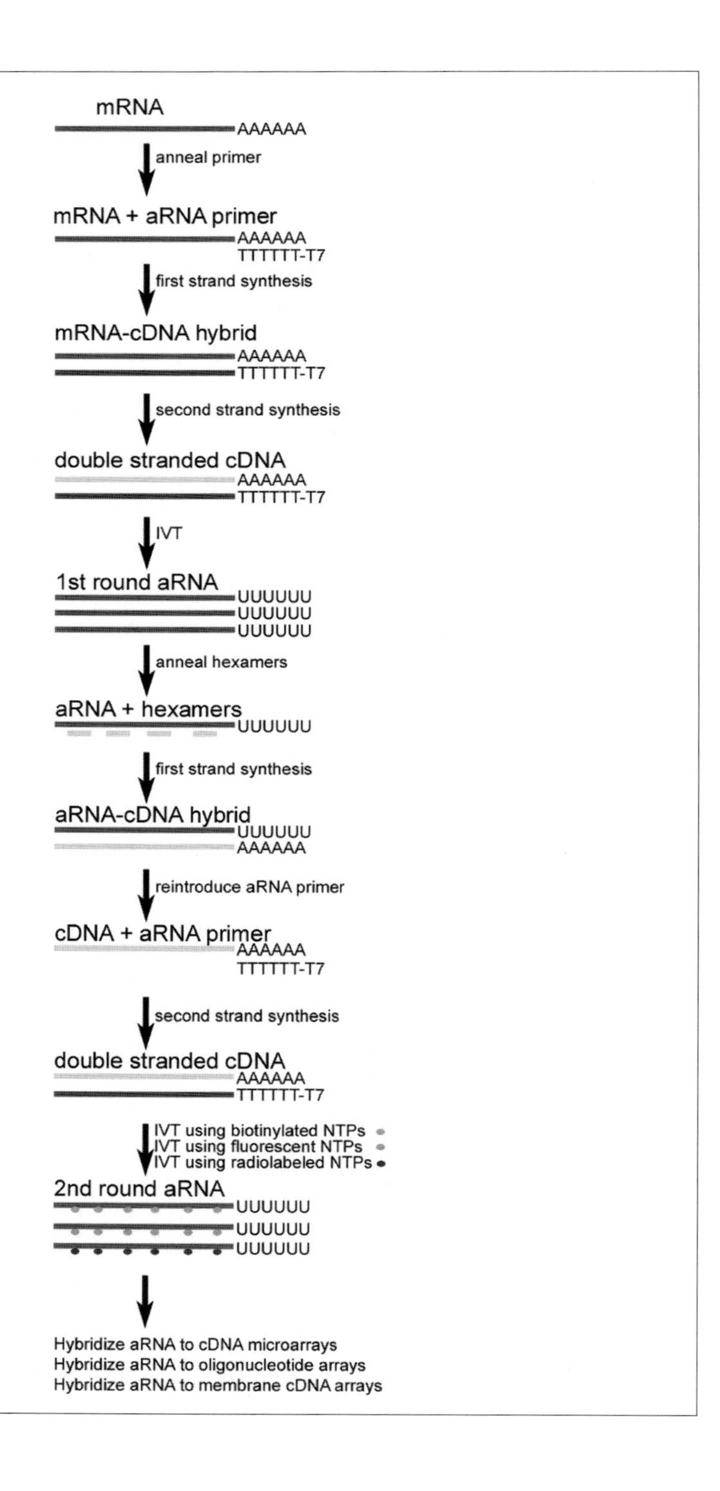
mRNA
AAAAAA
anneal primer
mRNA + aRNA primer
AAAAAA
TTTTTT-T7
first strand synthesis
mRNA-cDNA hybrid
AAAAAA
TTTTTT-T7
second strand synthesis
double stranded cDNA
AAAAAA
TTTTTT-T7
IVT
1st round aRNA
UUUUUU
UUUUUU
UUUUUU
anneal hexamers
aRNA + hexamers
UUUUUU
first strand synthesis
aRNA-cDNA hybrid
UUUUUU
AAAAAA
reintroduce aRNA primer
cDNA + aRNA primer
AAAAAA
TTTTTT-T7
second strand synthesis
double stranded cDNA
AAAAAA
TTTTTT-T7
IVT using biotinylated NTPs
IVT using fluorescent NTPs
IVT using radiolabeled NTPs
2nd round aRNA
UUUUUU
UUUUUU
UUUUUU
Hybridize aRNA to cDNA microarrays
Hybridize aRNA to oligonucleotide arrays
Hybridize aRNA to membrane cDNA arrays

RNA amplification strategies: TC RNA amplification

A novel linear RNA amplification procedure that utilizes a method of terminal continuation (TC) has been developed in our laboratory [39–41]. TC RNA amplification entails synthesizing first strand cDNA complementary to the input RNA, and IVT using the synthesized cDNA as template [39–41]. Second strand synthesis can be employed as part of the TC RNA amplification procedure, but is not requisite [42, 43]. Synthesis of first strand cDNA entails the use of two oligonucleotide primers, a poly d(T) primer and a TC primer. By providing a known sequence at the 3' region of first strand cDNA and a primer complementary to it, hairpin loops will not form. One round of amplification is sufficient for downstream genetic analyses [1, 7, 40, 44]. Transcription can be driven using a promoter sequence attached to either the 3' or 5' oligonucleotide primers. Therefore, transcript orientation can be in an antisense orientation (similar to conventional aRNA methods) when the bacteriophage promoter sequence is placed on the poly d(T) primer, or in a sense orientation when the promoter sequence is attached to the TC primer, depending on the design of the experimental paradigm (Fig. 2). Following TC RNA amplification, a large proportion of genes can be assessed for presence detection and by quantitative analysis as evidenced by bioanalysis and microarrays in cell culture preparations, animal model and human postmortem tissues [1, 45–50]. The threshold of detection of genes with low hybridization signal intensity is significantly increased with the TC RNA amplification method [40, 41], and background hybridization is significantly attenuated [41, 43]. Modifications of the TC RNA amplification technology are now being employed for the study and quantitation of microRNAs and related noncoding RNA species in tissue samples [1, 42].

Figure 1.
Schematic representation of the aRNA amplification. An oligo d(T)T7 primer is hybridized directly to poly(A) mRNAs, and a double stranded mRNA-cDNA hybrid is formed (first strand synthesis) by reverse transcribing the primed mRNAs with dNTPs and reverse transcriptase. Double stranded mRNA-cDNA hybrid is converted into double stranded cDNA, forming a functional T7 RNA polymerase promoter. A first round of aRNA synthesis occurs via *IVT using T7 RNA polymerase and NTPs. A second round of aRNA amplification occurs by annealing random hexamers to the newly formed aRNA, and performing first strand synthesis. The aRNA primer is then reintroduced, binding to the poly(A) sequence on the newly synthesized cDNA strand. A double stranded cDNA template is formed by second strand synthesis. A second round of aRNA products is produced by IVT using biotinylated, fluorescent, or radiolabeled NTPs. This figure is adapted from Ginsberg, 2005 [39] and Ginsberg and Mirnics, 2006 [43], and used with permission from the publishers.*

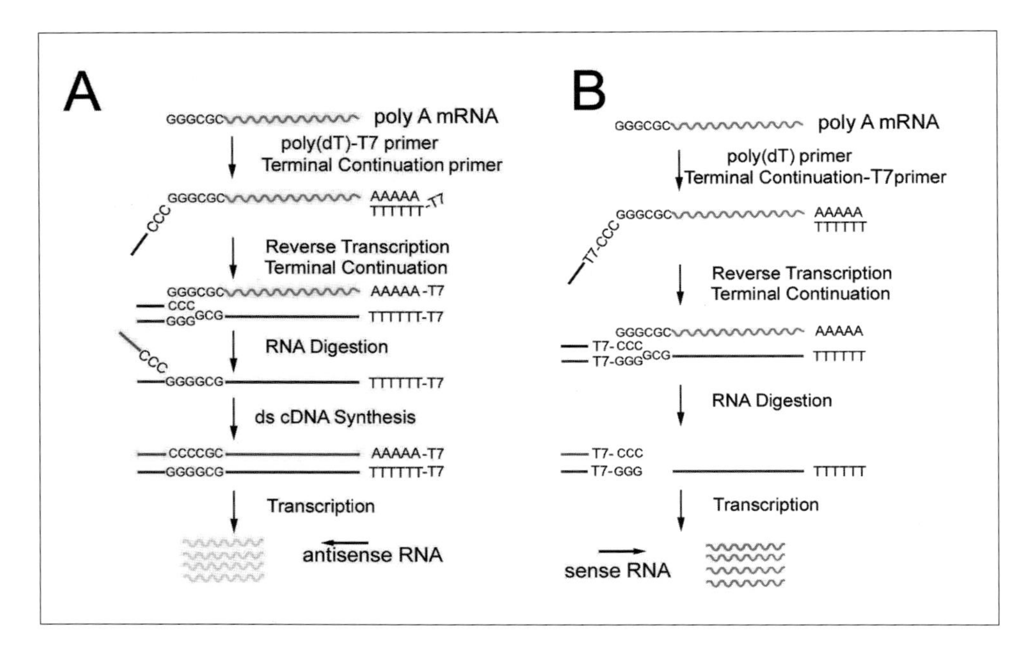

Figure 2
Overview of the TC RNA amplification method.
A. A TC primer (containing a T7 bacteriophage promoter sequence for sense orientation) and a poly d(T) primer are added to the mRNA population to be amplified (rippled line). First strand synthesis occurs as an mRNA-cDNA hybrid is formed following reverse transcription and TC of the oligonucleotide primers. After an RNase H digestion step to remove the original mRNA template strand, second strand synthesis is performed using Taq polymerase. The resultant double stranded product is utilized as template for IVT, yielding high fidelity, linear RNA amplification of antisense orientation (rippled lines).
B. Schematic illustrating the TC RNA amplification procedure amplifying RNA in the sense orientation (rippled lines) by placing the T7 RNA polymerase promoter sequence on the TC primer. Note that second strand synthesis is not required for sense TC RNA amplification. This figure is adapted from Che and Ginsberg, 2004 [40], and Ginsberg and Mirnics, 2006 [43] and used with permission from the publishers.

RNA preparation

RNA amplification techniques are useful with limiting amounts of input RNA. However, these procedures do not diminish the importance of careful sample handling, as small quantities of RNAs are highly vulnerable to RNase degradation [51]. An RNase-free environment is absolutely critical for successful RNA amplification procedures and downstream genetic analyses. Long term storage of RNAs (especially amplified RNAs) is not recommended and repeated freeze-thaw should be avoided. Furthermore, dubious RNA handling can complicate any troubleshooting process when problems arise. Feasible break points include following first strand synthesis, and immediately prior to IVT and the labeling procedure therein.

Our laboratory routinely purifies total cellular RNA (from regional dissections as well as LCM preparations) with Trizol reagent (Invitrogen). Trizol contains guanidine isothiocyanate and phenol. Tissues or single cells are directly submerged in Trizol reagent in a microfuge tube or homogenizing vessel. For whole paraffin embedded tissue sections, proteinase K digestion is included to ensure that cell membranes are disrupted and that RNAs are released from the tissues prior to adding Trizol reagent [6, 40, 52]. Trizol reagent inactivates RNases and therefore maintains the integrity of the RNA, while disrupting cells and dissolving cell components. RNAs purified from Trizol reagent are total cellular RNAs. mRNA can be affinity purified using oligo d(T) columns or beads. Kits are available to perform RNA (total and/or mRNA) extraction from *in vivo* and *in vitro* sources provided that a sufficient quantity of total RNA is available. Notably, the yield of mRNA from a single cell/population of cells captured by LCM may not withstand multiple purification steps. We have found that total RNA obtained from cells acquired *via* LCM is adequate for aRNA and TC RNA amplification [6, 7, 28, 29, 39].

In vitro transcription

IVT enables RNA synthesis through the activity of an RNA polymerase on a DNA template coupled to a bacteriophage transcription promoter sequence. In contrast to DNA synthesis, RNA synthesis does not need a primer, but a promoter sequence is required. To initiate RNA synthesis, a DNA-dependent RNA polymerase binds to a specific promoter DNA sequence that lies upstream of the transcription start site, and the strands of the cDNA separate in the vicinity of the transcription bubble which initiates RNA synthesis using the coding strand of the DNA as a template [53–55]. RNA polymerases used for IVT include T7, T3, and SP6 polymerases, named for the bacteriophages from which they were cloned. These aforementioned enzymes are single subunit RNA polymerases capable of transcribing complete genes without requiring additional proteins [56]. A specific and distinct bacteriophage RNA synthesis promoter is required for each polymerase, termed the T7, T3, and SP6 bacteriophage transcription promoter, respectively (Tab. 1). Although consensus sequences of all three promoters presumably contain a minimum sequence for activation (e.g., from –17 to +1; Tab. 1), this may not be enough sequence to initiate efficient RNA synthesis. For example, levels of IVT are virtually undetectable when a T7 promoter of the minimum consensus sequence composition is attached to a cDNA template [27]. In contrast, a T7 promoter of approximately 50 nucleotides initiated a robust IVT of both antisense RNA and sense RNA from TC

Table 1 - Depicted are representative bacteriophage transcription promoter sequences. The underlined sequences are consensus sequences of the respective promoters. Specificity of each promoter has been confirmed functionally. Notably, when a DNA vector contained all three phage promoters in tandem, each RNA polymerase was found to bind its own promoter and initiate transcription from a specific starting point without cross talk to other sequences.

Promoter	+1
T7	+1
TAATACGACTCACTATAG**G**GAGA	
T3	+1
AATTAACCCTCACTAAAGGGAGA	
SP6	+1
ATTTAGGTGACACTATAGAAGNG	

RNA amplification reactions [39, 40, 52]. Moreover, decreased efficiency of IVT is observed when a bacteriophage transcription promoter without a leader sequence is attached to a PCR-generated template [57].

RNA amplification: Hybridization to array platforms

Once an RNA amplification procedure is utilized to increase the input source of RNA species, biotinylated, fluorescent, or radiolabeled probes can be generated for subsequent hybridization to microarray platforms. Technical advances have fostered the development of high-density microarrays that allow for high throughput analysis of hundreds to thousands of genes simultaneously. Synthesis of cDNA microarrays entails adhering cDNAs or expressed sequence-tagged cDNAs (ESTs) to solid supports such as glass slides, plastic slides, or nylon membranes [58, 59]. A parallel technology uses photolithography to adhere oligonucleotides to array media, exemplified by the Affymetrix gene chip [60]. Gene expression using LCM acquired cells is assayed by harvesting total RNA or mRNA from the microdissected cells, amplifying the RNA with label incorporation, and hybridizing the labeled probes to the desired array platform. Arrays are washed to remove nonspecific background hybridization, and imaged using a laser scanner for biotinylated/fluorescently labeled probes and a phosphor imager for radioactively labeled probes. The specific signal intensity (minus background) of amplified RNA bound to each probe set (e.g., oligonucleotides or cDNAs/ESTs) is expressed as a ratio of the total hybridization signal intensity of the array, thereby minimizing variations due to differences in the specific activity of the probe and the absolute quantity of probe present. Gene expression data collected using single cells and/or homogeneous populations do not allow absolute quantitation of mRNA levels, but generate an expression profile of the relative changes in mRNA levels [6, 22, 30, 43, 45]. Relative changes in individual mRNAs are analyzed by univariate statistics (e.g., analysis of variance (ANOVA) with post-hoc Neumann-Keuls test) for individual comparisons [1, 31, 61]. Differential expression greater than approximately two-fold is accepted conventionally as relevant for further examination, although lower fold-difference levels can attain statistical significance and may be valuable as well [43, 62, 63]. Differentially expressed genes can be clustered into functional protein categories for multivariate coordinate gene expression analyses [64, 65]. Computational analysis is critical for optimal use of microarrays due to the enormous volume of data that is generated from a single probe. Additionally, access to relational databases is desirable, especially when evaluating hundreds of ESTs/probe sets that may/may not be linked to genes (and subsequent proteins) of known function.

Conclusions

Microarray analysis is being utilized to validate hypotheses and guide hypothesis-driven science rather than simply trawling for new genes. Thus, microarray analysis is based upon classes of transcripts relevant to the specific paradigms under investigation and to molecular and cellular mechanisms therein. There is great urgency to bridge basic science with translational studies to develop novel agents for discovery science. Prime examples include inflammation in the brain caused by a wide variety of disease states, ranging from age-related neurodegenerative disorders to traumatic brain injury to HIV-related neuropathology, where current therapies are not optimal and are often based on treating symptoms of the disease process. A functional genomics approach at the cellular level may identify novel agents that target selectively vulnerable cells (including, but not limited to neurons, glial cells, macrophages, and epithelial cells) during precise stages of disease pathogenesis. Ultimately, fingerprinting of selectively vulnerable cell types will provide a foundation to aid in ameliorating or preventing negative sequelae and mortality. The combination of single cell/population microdissection *via* LCM, RNA amplification, and microarray analysis enables high resolution, high throughput expression profiling of hundreds to thousands of genes simultaneously from a single cell or population. The next level of understanding of cellular and molecular mechanisms underlying normative function and the pathophysiology of disease lies in the ability to combine these aforementioned technologies with appropriate models to recapitulate the structure and connectivity of these systems *in vivo* and *in vitro*. There is a preponderance of evidence that complex biological processes are not governed by the action of a single isolated gene, but rather by the coordinate interactions of numerous genes, whose increased or diminished expression may play a role in normative function under physiological conditions and may contribute to the mechanisms underlying disease states under pathological conditions. Experimental paradigms such RNA molecular fingerprinting studies may help develop novel agents that target specific circuits or systems that are affected selectively and/or specifically during precise stages in disease pathogenesis. Therefore, this reduces the potential problems of drug interactions and unwanted side effects. In summary, T7-based RNA amplification procedures, notably the TC RNA amplification methodology, have broad applications for basic science and translational studies.

Acknowledgements
I am grateful for the technical and collaborative support of Melissa J. Alldred, Ph.D., Shaoli Che, M.D., Ph.D., Scott E. Counts, Ph.D., Irina Elarova, Shaona Fang, and Elliott J. Mufson, Ph.D. Support for this project comes from the NINDS (NS48447), NIA (AG09466, AG14449, AG17617), and Alzheimer's Association. I also acknowledge the families of the patients studied here who made this research possible.

References

1 Ginsberg SD (2007) Expression profile analysis of brain aging. In: DR Riddle (ed): *Brain Aging: Models, Methods and Mechanisms.* CRC Press, New York, 159–185
2 Ginsberg SD, Che S, Counts SE, Mufson EJ (2006) Single cell gene expression profiling in Alzheimer's disease. *NeuroRx* 3: 302–318
3 Galvin JE (2004) Neurodegenerative diseases: pathology and the advantage of single-cell profiling. *Neurochem Res* 29: 1041–1051
4 Phillips J, Eberwine JH (1996) Antisense RNA amplification: a linear amplification method for analyzing the mRNA population from single living cells. *Methods Enzymol (*Suppl) 10: 283–288
5 Sambrook J, Russell DW (2001) *Molecular cloning: a laboratory manual.* Third edition. Cold Spring Harbor Laboratory Press, Cold Spring Harbor
6 Ginsberg SD, Hemby SE, Mufson EJ, Martin LJ (2006) Cell and tissue microdissection in combination with genomic and proteomic applications. In: L Zaborszky, FG Wouterlood, JL Lanciego (eds): *Neuroanatomical Tract Tracing 3: Molecules, Neurons, and Systems.* Springer, New York, 109–141
7 Ginsberg SD, Che S (2004) Combined histochemical staining, RNA amplification, regional, and single cell analysis within the hippocampus. *Lab Invest* 84: 952–962
8 Su JM, Perlaky L, Li XN, Leung HC, Antalffy B, Armstrong D, Lau CC (2004) Comparison of ethanol *versus* formalin fixation on preservation of histology and RNA in laser capture microdissected brain tissues. *Brain Pathol* 14: 175–182
9 Lu L, Neff F, Dun Z, Hemmer B, Oertel WH, Schlegel J, Hartmann A (2004) Gene expression profiles derived from single cells in human postmortem brain. *Brain Res Brain Res Protoc* 13: 18–25

10 Goldsworthy SM, Stockton PS, Trempus CS, Foley JF, Maronpot RR (1999) Effects of fixation on RNA extraction and amplification from laser capture microdissected tissue. *Mol Carcinog* 25: 86–91

11 Ehrig T, Abdulkadir SA, Dintzis SM, Milbrandt J, Watson MA (2001) Quantitative amplification of genomic DNA from histological tissue sections after staining with nuclear dyes and laser capture microdissection. *J Mol Diagn* 3: 22–25

12 To MD, Done SJ, Redston M, Andrulis IL (1998) Analysis of mRNA from microdissected frozen tissue sections without RNA isolation. *Am J Pathol* 153: 47–51

13 Bonner RF, Emmert-Buck M, Cole K, Pohida T, Chuaqui R, Goldstein S, Liotta LA (1997) Laser capture microdissection: molecular analysis of tissue. *Science* 278: 1481–1483

14 Emmert-Buck MR, Bonner RF, Smith PD, Chuaqui RF, Zhuang Z, Goldstein SR, Weiss RA, Liotta LA (1996) Laser capture microdissection. *Science* 274: 998–1001

15 Kacharmina JE, Crino PB, Eberwine J (1999) Preparation of cDNA from single cells and subcellular regions. *Methods Enzymol* 303: 3–18

16 Polacek DC, Passerini AG, Shi C, Francesco NM, Manduchi E, Grant GR, Powell S, Bischof H, Winkler H, Stoeckert CJ Jr et al (2003) Fidelity and enhanced sensitivity of differential transcription profiles following linear amplification of nanogram amounts of endothelial mRNA. *Physiol Genomics* 13: 147–156

17 Feldman AL, Costouros NG, Wang E, Qian M, Marincola FM, Alexander HR, Libutti SK (2002) Advantages of mRNA amplification for microarray analysis. *Bio Techniques* 33: 906–914

18 Milligan JF, Groebe DR, Witherell GW, Uhlenbeck OC (1987) Oligoribonucleotide synthesis using T7 RNA polymerase and synthetic DNA templates. *Nucleic Acids Res* 15: 8783–8798

19 Melton DA, Krieg PA, Rebagliati MR, Maniatis T, Zinn K, Green MR (1984) Efficient *in vitro* synthesis of biologically active RNA and RNA hybridization probes from plasmids containing a bacteriophage SP6 promoter. *Nucleic Acids Res* 12: 7035–7056

20 Viale A, Li J, Tiesman J, Hester S, Massimi A, Griffin C, Grills G, Khitrov G, Lilley K, Knudtson K et al (2007) Big results from small samples: evaluation of amplification protocols for gene expression profiling. *J Biomol Tech* 18: 150–161

21 Ma C, Lyons-Weiler M, Liang W, LaFramboise W, Gilbertson JR, Becich MJ, Monzon FA (2006) *In vitro* transcription amplification and labeling methods contribute to the variability of gene expression profiling with DNA microarrays. *J Mol Diagn* 8: 183–192

22 Eberwine J, Kacharmina JE, Andrews C, Miyashiro K, McIntosh T, Becker K, Barrett T, Hinkle D, Dent G, Marciano P (2001) mRNA expression analysis of tissue sections and single cells. *J Neurosci* 21: 8310–8314

23 Eberwine J, Yeh H, Miyashiro K, Cao Y, Nair S, Finnell R, Zettel M, Coleman P (1992) Analysis of gene expression in single live neurons. *Proc Natl Acad Sci USA* 89: 3010–3014

24 VanGelder R, von Zastrow M, Yool A, Dement W, Barchas J, Eberwine J (1990) Ampli-

fied RNA (aRNA) synthesized from limited quantities of heterogeneous cDNA. *Proc Natl Acad Sci USA* 87: 1663–1667

25 Tecott LH, Barchas JD, Eberwine JH (1988) *In situ* transcription: specific synthesis of complementary DNA in fixed tissue sections. *Science* 240: 1661–1664

26 Madison RD, Robinson GA (1998) lRNA internal standards quantify sensitivity and amplification efficiency of mammalian gene expression profiling. *Bio Techniques* 25: 504–514

27 Ginsberg SD (2001) Gene expression profiling using single cell microdissection combined with cDNA microarrays. In: DH Geschwind (ed): *DNA Microarrays: The New Frontier in Gene Discovery and Gene Expression Analysis*. Society for Neuroscience Press, Washington, 61–70

28 Ginsberg SD, Crino PB, Hemby SE, Weingarten JA, Lee VM-Y, Eberwine JH, Trojanowski JQ (1999) Predominance of neuronal mRNAs in individual Alzheimer's disease senile plaques. *Ann Neurol* 45: 174–181

29 Ginsberg SD, Hemby SE, Lee VM-Y, Eberwine JH, Trojanowski JQ (2000) Expression profile of transcripts in Alzheimer's disease tangle-bearing CA1 neurons. *Ann Neurol* 48: 77–87

30 Hemby SE, Ginsberg SD, Brunk B, Arnold SE, Trojanowski JQ, Eberwine JH (2002) Gene expression profile for schizophrenia: discrete neuron transcription patterns in the entorhinal cortex. *Arch Gen Psychiat* 59: 631–640

31 Hemby SE, Trojanowski JQ, Ginsberg SD (2003) Neuron-specific age-related decreases in dopamine receptor subtype mRNAs. *J Comp Neurol* 456: 176–183

32 Xiang CC, Chen M, Ma L, Phan QN, Inman JM, Kozhich OA, Brownstein MJ (2003) A new strategy to amplify degraded RNA from small tissue samples for microarray studies. *Nucleic Acids Res* 31: E53

33 Wang E, Miller LD, Ohnmacht GA, Liu ET, Marincola FM (2000) High-fidelity mRNA amplification for gene profiling. *Nat Biotechnol* 18: 457–459

34 Luzzi V, Holtschlag V, Watson MA (2001) Expression profiling of ductal carcinoma *in situ* by laser capture microdissection and high density oligonucleotide arrays. *Am J Pathol* 158: 2005–2010

35 Matz M, Shagin D, Bogdanova E, Britanova O, Lukyanov S, Diatchenko L, Chenchik A (1999) Amplification of cDNA ends based on template-switching effect and step-out PCR. *Nucleic Acids Res* 27: 1558–1560

36 Brail LH, Jang A, Billia F, Iscove NN, Klamut HJ, Hill RP (1999) Gene expression in individual cells: analysis using global single cell reverse transcription polymerase chain reaction (GSC RT-PCR) *Mutat Res* 406: 45–54

37 Iscove NN, Barbara M, Gu M, Gibson M, Modi C, Winegarden N (2002) Representation is faithfully preserved in global cDNA amplified exponentially from sub-picogram quantities of mRNA. *Nat Biotechnol* 20: 940–943

38 Zhumabayeva B, Diatchenko L, Chenchik A, Siebert PD (2001) Use of SMART-generated cDNA for gene expression studies in multiple human tumors. *BioTechniques* 30: 158–163

39 Ginsberg SD (2005) RNA amplification strategies for small sample populations. *Methods* 37: 229–237

40 Che S, Ginsberg SD (2004) Amplification of transcripts using terminal continuation. *Lab Invest* 84: 131–137

41 Che S, Ginsberg SD (2006) RNA amplification methodologies. In: PA McNamara (ed): *Trends in RNA Research*. Nova Science Publishing, Hauppauge, 277–301

42 Che S, Alldred MJ, Ginsberg SD (2007) Microarray analysis using terminal continuation (TC) RNA amplification in human postmortem brain and animal models of neurodegeneration without second strand synthesis: implications for expression profiling and microRNA (miRNA) amplification. *Proc Soc Neurosci* 33: 795.716

43 Ginsberg SD, Mirnics K (2006) Functional genomic methodologies. *Prog Brain Res* 158: 15–40

44 Ginsberg SD, Che S (2005) Expression profile analysis within the human hippocampus: Comparison of CA1 and CA3 pyramidal neurons. *J Comp Neurol* 487: 107–118

45 Mufson EJ, Counts SE, Che S, Ginsberg SD (2006) Neuronal gene expression profiling: uncovering the molecular biology of neurodegenerative disease. *Prog Brain Res* 158: 197–222

46 White MM, Sheffer I, Teeter J, Apostolakis EM (2007) Hypothalamic progesterone receptor-A mediates gonadotropin surges, self priming and receptivity in estrogen-primed female mice. *J Mol Endocrinol* 38: 35–50

47 Ginsberg SD, Che S, Wuu J, Counts SE, Mufson EJ (2006) Down regulation of trk but not p75 gene expression in single cholinergic basal forebrain neurons mark the progression of Alzheimer's disease. *J Neurochem* 97: 475–487

48 Ginsberg SD, Che S, Counts SE, Mufson EJ (2006) Shift in the ratio of three-repeat tau and four-repeat tau mRNAs in individual cholinergic basal forebrain neurons in mild cognitive impairment and Alzheimer's disease. *J Neurochem* 96: 1401–1408

49 Counts SE, Chen EY, Che S, Ikonomovic MD, Wuu J, Ginsberg SD, DeKosky ST, Mufson EJ (2006) Galanin fiber hypertrophy within the cholinergic nucleus basalis during the progression of Alzheimer's disease. *Dement Geriatr Cogn Disord* 21: 205–214

50 Counts SE, He B, Che S, Ikonomovic MD, DeKosky ST, Ginsberg SD, Mufson EJ (2007) a7 nicotinic receptor up-regulation in cholinergic basal forebrain neurons in early stage Alzheimer's disease. *Arch Neurol* 64: 1771–1776

51 Blumberg DD (1987) Creating a ribonuclease-free environment. *Methods Enzymol* 152: 20–24

52 Ginsberg SD, Che S (2002) RNA amplification in brain tissues. *Neurochem Res* 27: 981–992

53 Severinov K (2001) T7 RNA polymerase transcription complex: what you see is not what you get. *Proc Natl Acad Sci USA* 98: 5–7

54 Kochetkov SN, Rusakova EE, Tunitskaya VL (1998) Recent studies of T7 RNA polymerase mechanism. *FEBS Lett* 440: 264–267

55 Skinner GM, Baumann CG, Quinn DM, Molloy JE, Hoggett JG (2004) Promoter bind-

ing, initiation, and elongation by bacteriophage T7 RNA polymerase. A single-molecule view of the transcription cycle. *J Biol Chem* 279: 3239–3244

56 Cheetham GM, Steitz TA (2000) Insights into transcription: structure and function of single-subunit DNA-dependent RNA polymerases. *Current Opinion in Structural Biology* 10: 117–123

57 Logel J, Dill D, Leonard S (1992) Synthesis of cRNA probes from PCR-generated DNA. *Bio Techniques* 13: 604–610

58 Schena M, Shalon D, Davis RW, Brown PO (1995) Quantitative monitoring of gene expression patterns with a complementary DNA microarray. *Science* 270: 467–470

59 Brown PO, Botstein D (1999) Exploring the new world of the genome with DNA microarrays. *Nat Genet* 21: 33–37

60 Lockhart DJ, Dong H, Byrne MC, Follettie MT, Gallo MV, Chee MS, Mittmann M, Wang C, Kobayashi M, Horton H et al (1996) Expression monitoring by hybridization to high density oligonucleotide arrays. *Nat Biotechnol* 14: 1675–1680

61 Mufson EJ, Counts SE, Ginsberg SD (2002) Single cell gene expression profiles of nucleus basalis cholinergic neurons in Alzheimer's disease. *Neurochem Res* 27: 1035–1048

62 Mirnics K, Levitt P, Lewis DA (2006) Critical appraisal of DNA microarrays in psychiatric genomics. *Biol Psychiatry* 60: 163–176

63 Unger T, Korade Z, Lazarov O, Terrano D, Sisodia SS, Mirnics K (2005) True and false discovery in DNA microarray experiments: transcriptome changes in the hippocampus of presenilin 1 mutant mice. *Methods* 37: 261–273

64 Aittokallio T, Kurki M, Nevalainen O, Nikula T, West A, Lahesmaa R (2003) Computational strategies for analyzing data in gene expression microarray experiments. *J Bioinform Comput Biol* 1: 541–586

65 Kotlyar M, Fuhrman S, Ableson A, Somogyi R (2002) Spearman correlation identifies statistically significant gene expression clusters in spinal cord development and injury. *Neurochem Res* 27: 1133–1140

Single and rare cell analysis – amplification methods

Amplification of cDNA from single or rare cells by global PCR (exponential amplification)

Christoph A. Klein and Claudia H. Hartmann

Division of Oncogenomics, Department of Pathology, University of Regensburg, Franz-Josef-Strauss-Allee 11, 93053 Regensburg, Germany

Abstract

In this chapter we provide a brief overview of protocols used to amplify mRNA isolated from single cells. We focus on PCR-based protocols and summarize our experiences in the use of amplified single cell cDNA for microarray hybridization.

Introduction

In recent years, important insights into the regulation or deregulation of cellular responses in various physiological and pathological states have been gathered through the generation of global gene expression profiles. The technique is now frequently used in many different areas of research including cancer, inflammation, and development [1–3].

Most studies applied microarray analyses to relatively large tissue samples or high cell numbers, because the first published gene expression protocols usually required large quantities of starting material, i.e. several micrograms of total RNA [4]. Recently, rare subpopulations or even individual cells have been placed into the centre of attention in many fields of scientific and clinical research. It has become clear that cellular heterogeneity and hierarchy within a multicellular organ and – in the case of cancer – the individual cancer cells as the units of mutation and selection, make higher cellular resolution necessary. Examples include single disseminated cancer cells that influence patient survival [5, 6], the development of different tissue types from stem and progenitor cells [7], or the many subsets of immune cells whose different roles are becoming ever clearer [8]. Consequently, the need for protocols that enable the analysis of rare or even single cells became evident.

Microarrays in Inflammation, edited by Andreas Bosio and Bernhard Gerstmayer

Studying the gene expression of single or rare cells

Single cells contain about 1–6 pg of mRNA with most transcripts being present in only 10–15 copies per cell [10]. Techniques to study the gene expression of single cells therefore need to be exquisitely sensitive, which is achieved by gene specific reverse transcription PCR that can be performed on a selected target mRNA. However, if the cells of interest are isolated and individual transcripts are to be directly quantified, the maximum number of PCR reactions is limited to about ten without prior amplification of the target cDNA. This limitation results from the fact that, in ten aliquots of the cellular cDNA, most transcripts will be present as single copy molecules. Thus, in any experimental setting diluting or splitting the sample to perform more than one PCR reaction will inevitably result in stochastic bias. Thus, for comprehensive transcriptional profiling of rare cells, gene specific approaches are less suited. For gene expression analysis on microarrays, which enable studying the expression of all known transcripts in parallel, several micrograms of RNA are usually required. Consequently, the single cell transcriptome needs to be amplified for this application. The major problem in single or rare cell gene expression profiling is therefore the undisturbed amplification of the cellular mRNA to amounts suitable for further analyses. To deal with this question, two individual and complementary methods have arisen. One is based on T7-based linear amplification of the starting RNA by *in vitro* transcription [11, 12], and the second on global PCR, which achieves an exponential increase of amplicons [13, 14]. Both approaches are summarized in Table 1 (see pp 98/99). A detailed discussion of T7-based linear amplification by Stephen D. Ginsberg can be found in a different section of this book. We will therefore concentrate on global PCR-based exponential protocols.

Exponential amplification of single cell mRNA by global PCR

Table 1 gives an overview of the published protocols. The most widely used methods include the so-called Brady procedure [15] and SCAGE [13, 16]. Both have certain criteria in common:

1. Amplification is based on the PCR technique. Thus, on the one hand, target sequences are amplified exponentially and, on the other hand, cDNAs that can reach lengths of several thousand base pairs must be reduced to a size suitable for PCR. This can be achieved either by limiting the conditions during cDNA synthesis (Brady) or by fragmentation during mRNA priming with random oligonucleotides (SCAGE) while maintaining optimal conditions for all enzymatic steps.
2. To amplify all cellular mRNAs regardless of their sequence, primer binding sites must be introduced into the target sequence. One primer binding site is gener-

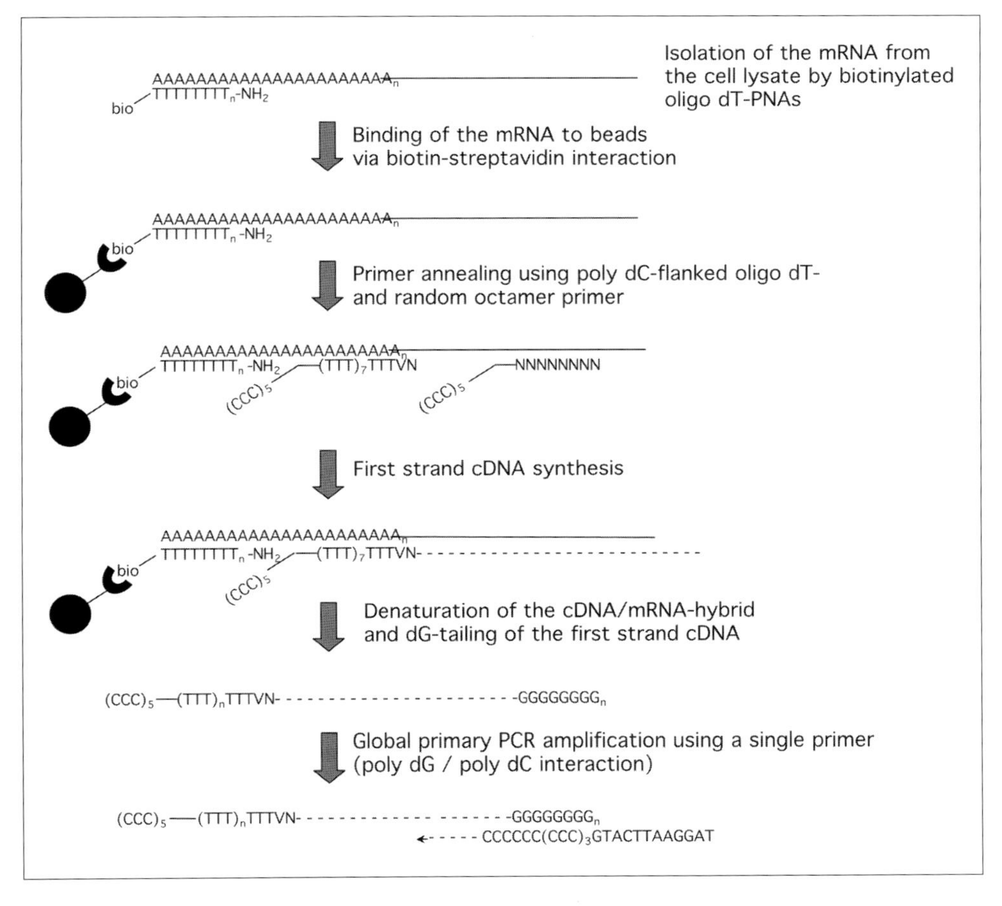

Figure 1
Schematic overview of the exponential amplification of cDNA using SCAGE [16]. The mRNA is selected from the cell lysate by biotinylated oligo-dT PNAs (Peptide nucleic acids) in combination with streptavidin beads. To fragment the nucleic acids to a size suitable for PCR amplification, random oligos are used during cDNA synthesis together with a T24-cDNA synthesis primer. After denaturing the cDNA/mRNA hybrids, a poly G-tailing reaction is performed. The resulting poly C/polyG stretches serve as uniform primer binding sites and allow the amplification of all cDNAs regardless of their sequence.

ated *via* the primer during first strand cDNA synthesis. For the second primer binding site, a tailing reaction is most often performed. The tailing can either use Adenosin (Brady) or Guanosin (SCAGE) as substrate. The following exponential amplification then uses a complementary poly-dT or poly-dC primer, respectively (Fig. 1).

Table 1 - Methods for amplification of limited amounts of mRNA

Method	Principle	Advantages	Limitations	Refs
T7-based amplification (Eberwine)	Linear amplification by T7-RNA polymerase	Widely used, well documented, kits available	Sensitivity for comprehensive single-cell gene expression analysis not convincingly demonstrated	[12, 28]
PCR-based amplification (Brady)	Exponential amplification using dT/dA primers	Relatively simple protocol, widely used Higher amplification efficiency than linear methods	cDNA synthesis under limiting conditions Limited sensitivity due to Oligo-dT/dA primer in PCR	[14, 15, 24]
PCR-based amplification (Kurimoto)	Exponential amplification using dT/dA primers	Higher amplification efficiency than linear methods Higher sensitivity than Brady (Two rounds of PCR plus T7 amplification) Applicable to oligonucleotide arrays using T7 technology	cDNA synthesis under limiting conditions Limited sensitivity due to Oligo-dT/dA primer in PCR complex protocol	[23]
PCR-based amplification (SCAGE)	Exponential amplification using dC/dG primers	Relatively simple protocol, robust, higher amplification efficiency compared to linear methods and dT/dA-PCR protocols Can be combined with SCOMP for analysis of genomic DNA from same cell (Table 1)	Limited sensitivity for very low abundant transcripts in single murine cells	[13]

Table 1 (continued)

Method	Principle	Advantages	Limitations	Refs
PCR-based amplification (SCAGE)	Exponential amplification using dC/dG primers, optimized mRNA extraction	Improved mRNA isolation and cDNA synthesis compared to previous SCAGE protocol Demonstrated sensitivity for rare transcripts in murine cells Applicable to oligonucleotide arrays Can be combined with SCOMP for analysis of genomic DNA from same cell (Table 1)	Use of T7-based hybridisation techniques and Affymetrix platforms only after modification of the amplification products	[16]
TALPAT	Combination of T7- and PCR-amplification	Higher amplification efficiency compared to T7-based linear amplification alone Applicable to oligonucleotide arrays using T7 technology	Sensitivity for single cells not convincingly demonstrated multistep protocol	[29]
RACE	Amplification of 3′-sequences of RNAs	Simple, fast	Mostly 3′-UTR amplified Currently limited to 40 genes	[30]

The PCR based methods are believed to be less complicated and more sensitive as compared to T7-polymerase based protocols, while it is often assumed that only linear amplification methods preserve the transcript ratios quantitatively. However, we were able to demonstrate by quantitative PCR that our PCR-based method preserves the relative transcript ratios [16]. For the T7-method [12], it is still unclear whether the original proportions of mRNAs are preserved, because a sequence-specific bias was observed during amplification at least in one of three studies [17–19]. For PCR-based approaches, it has been proposed that the amplification reaction should be stopped before reaching the plateau phase or that certain sequences are amplified preferentially [20–22]; however, neither we nor the group of Norman Iscove observed such an effect [14]. When we investigated single cells, quantitative PCR results demonstrated that relative transcript ratios were not altered when the template cDNA was taken from the plateau-phase of the amplification PCR. In addition, the correlation coefficients of the hybridization results relative to non-amplified total RNA dropped in all cases when cycling was stopped at the exponential phase of amplification, as compared to the samples taken from the plateau phase. This was probably due to a higher coefficient of variability in the presence of less template cDNA for labelling and hybridization. We therefore do not recommend stopping the primary amplification too early.

The protocol by Brady and Iscove

While no fully validated method for comprehensive transcriptional profiling of single cells has been published using linear amplification, several PCR-based protocols for single cell gene expression profiling on large-scale arrays have been established [16, 23, 24].

Tietjen and coworkers used the perhaps most frequently applied protocol for single cell amplification by Brady and Iscove [15] and hybridized the PCR products onto an Affymetrix platform. However, several aspects of this study remain unclear, such as the control for efficient and specific binding of the DNA to the short Affymetrix oligos that preferentially bind RNA; also, no comprehensive gene expression evaluation was shown. In principle, the protocol suffers from the limitations of the original Brady protocol which combines conditions of limited processivity for the enzymes used in cDNA synthesis and primers of imperfect amplification efficiency (see below and supplemental information in [13]). The protocol by Kurimoto et al. consists of two rounds of PCR amplification. During the second PCR, a T7-promoter sequence is introduced into the cDNA fragments to allow for the subsequent T7 based amplification, labelling, and hybridisation of the PCR products. Using qPCR, the authors demonstrated small inter-experimental variations of diluted total RNA, especially for genes with higher expression (i.e. 20 copies per cell). Since reliability and reproducibility were only assessed with diluted RNA, the efficiency for amplifying transcripts from single cells using this protocol, and thus its sensitivity, are unknown. The method was applied to morphologically indistinguishable single cells from mouse blastocysts and it was found that these cells consist of two individual cell populations (primitive endoderm- and epiblast-like) by embryonic day 3.5.

The SCAGE protocol

The already mentioned Brady procedure was also the starting point of our own modified version (called SCAGE for single cell analysis of gene expression [13]. We observed that the applied conditions of limited processivity for reverse-transcriptase during cDNA synthesis (e.g. low concentration of cDNA synthesis primers and nucleotides and short reaction time) and the poly-T primer for the PCR amplification severely reduced sensitivity and we developed a protocol to overcome these shortcomings [13, 25]. It was based on isolation of the cellular mRNA by oligo-dT beads, cDNA synthesis using random octamer and oligo-$dT_{(15)}$ primers, poly-dG tailing, and PCR amplification using a single primer under very stringent conditions adequate for CG-rich sequences [13]. The solid-phase capturing of the mRNA enabled the depletion of the cDNA from non-incorporated cDNA synthesis primers and abundant dNTPs, thus avoiding the subsequent amplification of tailed primers or inefficient tailing due to micromolar concentrations of 3'-ends. Together with the introduction of random primers (resulting in a fragmentation of the cDNA to a size amplifiable by PCR) the protocol then enabled applying optimal conditions for all enzymatic reactions such as high concentrations of dNTPs, primer, and enzyme, all of which were individually shown to contribute to increased sensitivity [13]. Using random primers for fragmentation of first strand cDNA has additional advantages. As these random oligonucleotides also contain the flanking region that serves as a primer binding site for the global PCR, 5'-regions of the mRNA are also reverse transcribed and amplified. Otherwise, these sequences, which may also be of interest, are lost. Second, using random hexa-, octa- or pentameres allows for the regulation of cDNA length. Our current protocol uses random octameres, resulting in an average cDNA fragment of 500 bp. But due to the use of the random primer, an initial removal of rRNA, tRNA, and genomic DNA is essential, and is achieved by the selection of the mRNA *via* oligo dT-beads in the very beginning. The greatest increase in sensitivity (100 fold) over the original Brady procedure was achieved by the use of poly-G tailing instead of poly-A tailing and the subsequent use of a single poly-C containing primer [13]. Making all sequences equally CG-rich apparently reduces sequence-dependent variation of amplification efficiency and maintains the relative transcript ratios.

However, even this protocol did not allow analysis of single cells with extremely low mRNA content, e.g. mouse or stem cells. We therefore further improved the method for genome-wide gene expression profiling of single cells on large-scale oligonucleotide microarrays suited for all cell types alike. This was recently achieved by modification of our original protocol. The process of isolating and amplifying cellular mRNA was improved to a great extent by introducing two modifications. First, application of biotinylated poly-T peptide nucleic acids (PNAs) bound to streptavidin beads instead of oligo-dT beads for the isolation of mRNA significantly increased yields. It appears that rare transcripts are isolated with much higher probability by PNAs because of their much higher affinity to the poly-A tail of the mRNA as compared to oligo-dT beads. Second, adding a longer oligo-$dT_{(24)}$-containing cDNA synthesis primer to the random octamer further improved results. The size of the cDNA is again restricted by usage of random primers during cDNA synthesis to a size optimal for PCR. The first primer-binding site (a poly dC stretch) for PCR is a flanking 5' region in all cDNA primers that are elongated during reverse transcription. The second poly-dG containing primer binding site is generated by a tailing reaction, enabling amplification of all fragments with a single poly-dC primer, as before. The quantitative nature of the amplification was demonstrated by qPCR experiments comparing the expression levels of specific genes from amplified and non-amplified single cells [16]. Moreover, a quantitative analysis of transcripts from single heart cells isolated from young *versus* old animals identified loss of transcriptional control of gene expression as a mechanism that is closely associated with ageing [26].

The modified protocol now enables retrieval of the correct histogenetic origin of cells and monitoring of cellular differentiation and pathway activation in individual cells [16]. On microarrays that comprise 17000 transcripts, epithelial cells, hematopoietic stem cells, immature and mature dendritic cells were all correctly classified, providing the most comprehensive gene expression study on large-scale arrays so far. Maturation of dendritic cells after stimulation with LPS were monitored in single cells by analysis of the Tlr4 pathway that is activated after LPS stimulation.

For hybridization of amplified single cell cDNA on large-scale microarrays containing 17000 oligonucleotides of 70 nt length, we use a secondary amplification in the presence of labelled nucleotides to label the PCR products for subsequent microarray analysis. Upon hybridisation on microarrays, we noted that contaminating bacterial sequences in the amplified sample interferes with reliable measurements if the array contains bacterial sequences as well. Such sequences originate from the various recombinant enzyme preparations (mostly DNA binding enzymes), which are used for primary amplification and are inevitably co-amplified and labelled, and will result in false-positive hybridisation signals if complementary sequences are also present on the array. Thus, using arrays made of cloned cDNAs was often unsuccessful. In contrast, oligonucleotide arrays are free from bacterial or plasmid sequences and, although hybridisation conditions for single cells differ from standard procedures for total RNA, robust experimental conditions could be identified. Sensitivity and specificity of hybridization techniques are controlled to a large extent by the stringency of the post-hybridization washing procedure. We therefore established a single cell hybridization and washing protocol that may differ from standard procedures.

Our data indicate that with this protocol, single cells express transcripts binding to 25–30% of all genes on the array, suggesting that low abundant transcripts are also detected with high reliability. The correlation coefficient of technical replicates for both amplification and hybridisation was found to range from 0.88 to 0.91, similar to what has been observed for samples from several microgram of total RNA. Also, validation of the array experiments by qPCR showed high concordance between the two methods and demonstrated validity of the hybridization results both with regard to qualitative and quantitative signal intensities. The observed differences between array and PCR results for single cells were similar to those published for conventional gene expression profiling studies [27].

In summary, PCR based amplification methods for whole genome amplification have demonstrated their suitability for quantitative assessment of multiple transcript numbers both by qPCR and global gene expression profiling. The published methods are well validated and have been used to gain insight into various scientific questions. It is anticipated that they will further prove their usefulness whenever rare cell populations are being studied.

References

1 Granucci F et al (2001) Inducible IL-2 production by dendritic cells revealed by global gene expression analysis. *Nat Immunol* 2(9): 882–8

2 van't Veer LJ et al (2002) Gene expression profiling predicts clinical outcome of breast cancer. *Nature* 415(6871): 530–6

3 Golub TR et al (1999) Molecular classification of cancer: class discovery and class prediction by gene expression monitoring. *Science* 286(5439): 531–7

4 Mahadevappa M, Warrington JA (1999) A high-density probe array sample preparation method using 10- to 100-fold fewer cells. *Nat Biotechnol* 17(11): 1134–6

5 Braun S et al (2005) A pooled analysis of bone marrow micrometastasis in breast cancer. *N Engl J Med* 353(8): 793–802

6 Janni W et al (2000) Prognostic significance of an increased number of micrometastatic tumor cells in the bone marrow of patients with first recurrence of breast carcinoma. *Cancer* 88(10): 2252–9

7 Reya T et al (2001) Stem cells, cancer, and cancer stem cells. *Nature* 414(6859): 105–11

8 Murphy KM, Travers P, Walport M (2008) *Immunobiology*. Taylor & Francis

9 Klein CA et al (1999) Comparative genomic hybridization, loss of heterozygosity, and DNA sequence analysis of single cells. *Proc Natl Acad Sci USA* 96(8): 4494–9

10 Alberts B et al (2002) *Molecular Biology of the Cell*. 4th edition ed. New York: Garland Publishing

11 Baugh LR et al (2001) Quantitative analysis of mRNA amplification by *in vitro* transcription. *Nucleic Acids Res* 29(5): E29

12 Van Gelder RN et al (1990) Amplified RNA synthesized from limited quantities of heterogeneous cDNA. *Proc Natl Acad Sci USA* 87(5): 1663–7

13 Klein CA et al (2002) Combined transcriptome and genome analysis of single micrometastatic cells. *Nat Biotechnol* 20(4): 387–92

14 Iscove NN et al (2002) Representation is faithfully preserved in global cDNA amplified exponentially from sub-picogram quantities of mRNA. *Nat Biotechnol* 20(9): 940–3

15 Brady G et al (1993) Construction of cDNA libraries from single cells. *Methods Enzymol* 225: 611–23

16 Hartmann CH et al (2006) Gene expression profiling of single cells on large-scale oligonucleotide arrays. *Nucleic Acids Res* 34(21): e143

17 Feldman AL et al (2002) Advantages of mRNA amplification for microarray analysis. *Biotechniques* 33(4): 906–12, 914

18 Heil SG et al (2003) Gene-specific monitoring of T7-based RNA amplification by real-time quantitative PCR. *Biotechniques* 35(3): 502–4, 506–8

19 Li J et al (2003) RNA amplification, fidelity and reproducibility of expression profiling. *C R Biol* 326(10–11): 1021–30

20 Dixon AK et al (2000) Gene-expression analysis at the single-cell level. *Trends Pharmacol Sci*, 2000. 21(2): p. 65–70

21 Freeman TC et al (1999) Analysis of gene expression in single cells. *Curr Opin Biotechnol* 10(6): 579–82

22 Saghizadeh M et al (2003) Evaluation of techniques using amplified nucleic acid probes for gene expression profiling. *Biomol Eng* 20(3): 97–106

23 Kurimoto K et al (2006) An improved single-cell cDNA amplification method for efficient high-density oligonucleotide microarray analysis. *Nucleic Acids Res* 34(5): e42

24 Tietjen I et al (2003) Single-cell transcriptional analysis of neuronal progenitors. *Neuron* 38(2): 161–75

25 Klein CA et al (2002) The hematopoietic system-specific minor histocompatibility antigen HA-1 shows aberrant expression in epithelial cancer cells. *J Exp Med* 196(3): 359–68

26 Bahar R et al (2006) Increased cell-to-cell variation in gene expression in ageing mouse heart. *Nature* 441(7096): 1011–4

27 Park PJ et al (2004) Current issues for DNA microarrays: platform comparison, double linear amplification, and universal RNA reference. *J Biotechnol* 112(3): 225–45

28 Eberwine J et al (1992) Analysis of gene expression in single live neurons. *Proc Natl Acad Sci USA* 89(7): 3010–4

29 Aoyagi K et al (2003) A faithful method for PCR-mediated global mRNA amplification and its integration into microarray analysis on laser-captured cells. *Biochem Biophys Res Commun* 300(4): 915–20

30 Fink L et al (2002) cDNA array hybridization after laser-assisted microdissection from nonneoplastic tissue. *Am J Pathol* 160(1): 81–90

Selected applications of microarrays in inflammation research

Gene expression patterns in asthma

Kenji Izuhara[1], Sachiko Kanaji[1], Shoichiro Ohta[1], Hiroshi Shiraishi[1], Kazuhiko Arima[1], Noriko Yuyama[2]

[1]Division of Medical Biochemistry, Department of Biomolecular Sciences, Saga Medical School, Saga, 849-8501, Japan; [2]Genox Research, Inc., Tokyo, 154-0004, Japan

Abstract

Bronchial asthma is a complicated and diverse disorder affected by genetic and environmental factors, with Th2-type inflammation dominant in its pathogenesis. However, the underlying molecular mechanism of bronchial asthma is still poorly understood. Microarray technology, now one of the most powerful tools for functional genomics, has been used in several trials to dissect the pathogenesis of bronchial asthma, providing some novel pathogenic mechanisms as well as information about gene-expression profiling. This article describes the recent outcomes of microarray analyses applied to bronchial tissues of asthma patients or asthma animal models and cultured cells related to the biological events in bronchial asthma. This information could be relevant for finding drug targets or biomarkers for bronchial asthma.

Introduction

It is estimated that 300 million people suffer from bronchial asthma, and the number of patients is still increasing [1]. Moreover, the medical cost for treating asthma patients is huge and on the increase. Therefore, it is important socially as well as medically to clarify the pathogenesis of bronchial asthma and to establish more useful strategies to overcome it.

Bronchial asthma is a complicated and diverse disorder affected by genetic and environmental factors. Inhalation of ubiquitous allergens causes Th2-type dominant inflammation in lungs, leading to bronchial asthma [2, 3]. The importance of Th2-type responses in the pathogenesis of bronchial asthma has been confirmed by analyses of mice models, expressed cytokine profiles in asthma patients, and the existence of variants susceptible to bronchial asthma in Th2-type cytokine-signaling molecules [4]. Th2-type responses induce production of various effectors contributing to the onset or the exacerbation of bronchial asthma. Furthermore, inflammation processes other than Th2-type responses also modify the complicated pathogenesis of bronchial asthma. Therefore, it is important to comprehensively identify the molecules involved in the cascades of the inflammatory processes in asthma and

Microarrays in Inflammation, edited by Andreas Bosio and Bernhard Gerstmayer

to elucidate the network interaction between these molecules and the inflammatory cells or resident cells in the lungs.

Microarray technology is now one of the most powerful tools for functional genomics. It has been used in several trials to dissect the pathogenesis of bronchial asthma, providing some novel pathogenic mechanisms of bronchial asthma as well as information about gene expression profiling. Information on gene expression profiling should be relevant for finding drug targets or biomarkers for bronchial asthma. This article describes the recent outcomes of microarray analyses applied to bronchial tissues of asthma patients or asthma model animals and cultured cells related to the biological events in bronchial asthma.

Application of microarray technology to human bronchial tissues

The analysis of gene expression in bronchial tissues of asthma patients is the most direct way to apply microarray technology to bronchial asthma. In our survey, three groups reported gene expression profiles in bronchial tissues of asthma patients using microarray analysis [5–7]. These analyses are composed of the following comparisons: (1) bronchial biopsies from healthy controls and asthma patients before and after inhaled corticotherapy [5]; (2) airway epithelium in asthmatic patients before and after segmental allergen challenge [6]; and (3) nasal respiratory epithelial cells from healthy children and those with stable and exacerbated asthma [7]. They demonstrate different expression patterns in each comparison, which may be clinically useful. However, these reports did not address how the specific product of the identified genes was involved in the pathogenesis of bronchial asthma.

Application of microarray technology to animal models

Because the genetic and environmental backgrounds of asthma patients are diverse and the condition of clinical course and treatment depends on the patients, an approach using human tissues sometimes fails to reach stable and clear conclusions. However, the application of microarray to animal models can help overcome this problem. Furthermore, using particular genetically engineered mice enables us to dissect gene expression changes correlated with specific mediators, cell types, or pathological features. Several trials using animal models have been performed, most of which are mouse-based with one monkey-based, demonstrating several novel underlying mechanisms of bronchial asthma [8–12].

Zimmermann et al. tried to identify inducible genes in lung tissues derived from ovalbumin- or *Aspergillus*-inducible asthmatic mice [8]. The overlapping upregulated genes in ovalbumin- or *Aspergillus*-inducible asthmatic mice contained arginase I, arginase II, and cationic amino acid transporter 2 (CAT). These three molecules are involved in the uptake and metabolism of arginine into ornithine. The precise role of activation of the arginase pathway in bronchial asthma is unclear. However, polyamines, metabolites of ornithine, may interact with macromolecules including RNA or DNA, thereby regulating cell growth, division, and differentiation [13]. Alternatively, because ornithine is a precursor of proline, a major component of collagen, the arginase pathway may play an important role in fibrosis in bronchial asthma.

Karp et al. designed a unique experiment to try to identify a susceptibility factor to airway hyperreactivity (AHR) using two different strains of mice, A/J and C3H/HeJ strains, highly susceptible and resistant to allergen-induced AHR, respectively [9]. They focused on complement 5 (C5) among differently expressed genes, because the *C5* gene located close to a locus correlated with allergen-induced bronchial hyperresponsiveness. A/J mice, but not C3H/HeJ mice, had a 2-bp deletion in a 5' exon of the *C5* gene that rendered them deficient in C5 mRNA and protein production. Furthermore, they found that C5 induced IL-12 production in monocytes. These results indicated that C5 is involved in determining susceptibility to bronchial asthma by inducing IL-12 production, which counterbalances the Th2-type immune responses.

Munitz et al. found that CD48 was induced in lung tissues of ovalbumin-challenged mice, independently of STAT6, IL-4, and IL-13 [10]. CD48 is a glycosylphosphatidylinositol-anchored protein belonging to the CD2 subfamily. It is a low-affinity and high-affinity ligand for CD2 and 2B4, respectively, and is involved in cell-cell interaction followed by activation of various CD48-bearing cells. They showed that expression of CD48 was upregulated in various hematopoietic cells, including eosinophils, lymphocytes, macrophages and NK cells in asthmatic lungs, and that blocking of the CD48 pathway attenuated lung inflammation.

Kuperman et al. applied a microarray approach to three kinds of asthmatic mice models: (1) wild-type mice exposed to ovalbumin (Ova); (2) mice expressing IL-13 in epithelium (tg-IL-13); and (3) mice expressing IL-13 in epithelium with STAT6 expression limited to non-ciliated airway epithelial cells (IL-13/Epi) [11]. These three types of mice showed different phenotypes: mucus production and enhanced AHR were common features in all the mice, whereas atopy and fibrosis/emphysema were specific features for Ova and tg-IL-13, respectively. They grouped the identified genes based on the combination of increased expression in each kind of mouse. These data were useful to elucidate and monitor specific characteristics of bronchial asthma by focusing attention on a limited number of identified genes.

A DNAX group tried to identify *Ascaris*- or IL-4-inducible genes in lung tissues derived from asthmatic monkeys by the microarray approach [12]. The allergic cynomolgus monkey (*Macaca fascicularis*), which they used for the analyses, has a natural hypersensitivity to the antigen of the nematode *Ascaris suum*. They confirmed upregulation of several chemokines such as MCP-1, MCP-3, and eotaxin. To our knowledge, this is the only subject for which a microarray approach was applied to a non-human primate asthma model.

Application of microarray technology to cultured cells

In contrast to the analyses of human specimens or model animals, application of the microarray approach to cultured cells has an advantage in analyzing expression profiles focusing on specific cells or conditions. Resident cells in bronchial tissues such as epithelial cells, fibroblasts, and smooth muscle cells [14–16] or inflammatory cells such as mast cells [17] have been subjected to microarray analysis.

Lee et al. demonstrated the gene lists induced by IL-13 in bronchial epithelial cells, smooth muscle cells, and lung fibroblasts [14]. IL-13 acts as a central mediator of bronchial asthma [18–20]. Very few genes overlapped in expression profiles in three kinds of cells.

Chu et al. used the microarray approach to analyze the effects of compression on human bronchial epithelial cells [15]. Compression is a mechanical stress experienced by airway epithelial cells as a result of bronchoconstriction. They found that compressed epithelial cells upregulated expression of plasminogen activating system-related genes, such as urokinase plasminogen activator, urokinase plasminogen activator receptor, plasminogen activator inhibitor-1 (PAI-1), and tissue plasminogen activator, indicating that urokinase-dependent plasminogen generation could be accelerated in bronchial asthma.

Cho et al. also showed the involvement of the plasminogen pathway in the pathogenesis of bronchial asthma by applying microarray technology to a human mast cell line, HMC-1, activated by phorbol ester and calcium ionophore [17]. They found that expression of PAI-1 was upregulated in these cells. Because PAI-1 inhibits the plasminogen activator converting plasminogen to plasmin, which enhances proteolytic degradation of the extracellular matrix, activated mast cells could play an important role in airway remodeling by secreting PAI-1.

We tried to identify the inducible genes in bronchial epithelial cells by either IL-13 or IL-4, sharing the same receptor composed of the IL-4 receptor α chain and the IL-13 receptor $\alpha1$ chain. We found that expression of 12 genes was reproducibly enhanced by both IL-4 and IL-13 [16]. Among the gene products of the identified genes, we performed functional analyses of periostin, examining the correlation with the pathogenesis of bronchial asthma [21]. Periostin has four fasciclin 1 domains, which are characteristic of a family of adhesion molecules, and it is thereby assumed to be an adhesion molecule. However, nothing had been known about the correlation between periostin and bronchial asthma. Upon stimulation of IL-13, periostin was secreted from lung fibroblasts. Histochemical analysis showed that periostin was localized in thickened basement membranes of bronchial tissues in asthma patients, but no deposition of periostin was observed in normal donors. Deposition of periostin in fibrotic lesions of an ovalbumin-inhaled "chronic asthma mouse model" was mostly diminished in IL-13 knockout mice, showing that induction of periostin was located downstream of IL-13 signals. Furthermore, we found that periostin bound to other matrix proteins composing fibrosis of bronchial

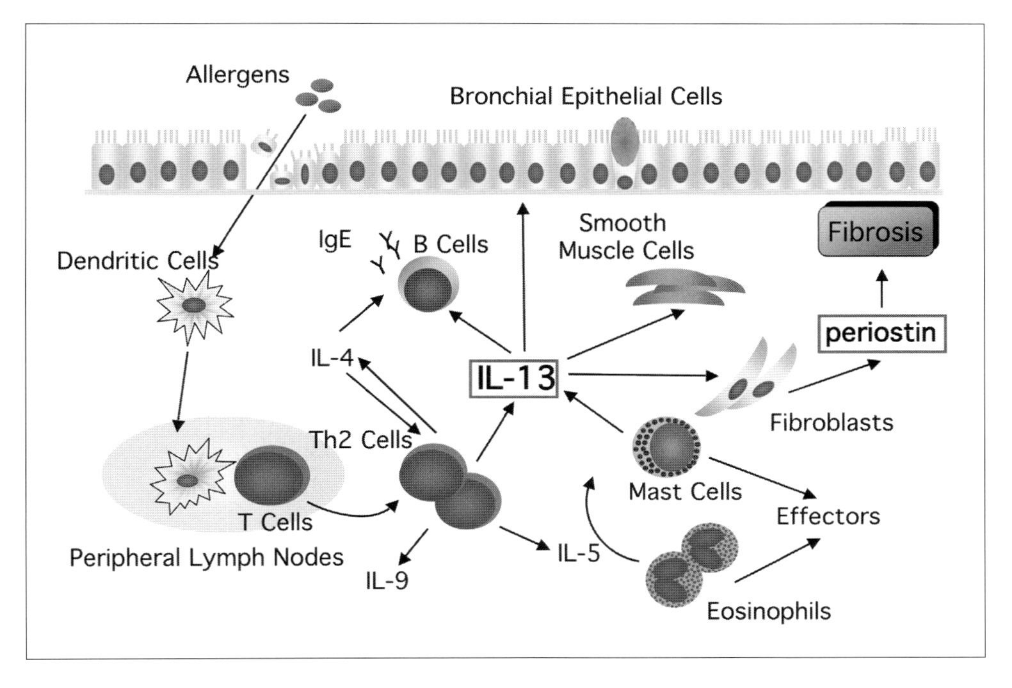

Figure 1
Role of periostin in the pathogenesis of bronchial asthma
The role of periostin in the pathogenesis of bronchial asthma is depicted. Invaded allergens induce Th2-type responses in the host. Th-2 cells secrete Th2-type cytokines, such as IL-4, 5, 9, and 13. IL-13 acts on fibroblasts and induces periostin. Periostin contributes to fibrosis by binding to other matrix proteins.

asthma, such as tenascin-C, fibronectin, and collagen V. Taken together, these findings indicate that periostin is a novel component of fibrosis of bronchial asthma and contributes to fibrosis by binding to other matrix proteins (Fig. 1).

We then analyzed the squamous cell carcinoma antigen 1 (SCCA1) and SCCA2, two other gene products among the 12 upregulated identified genes. SCCA1 and SCCA2 are members of the ovalbumin serpin family; however, the physiological targets of these protease inhibitors were uncertain [22]. We found that SCCA1 and SCCA2 inhibited the cysteine protease activities of parasite-derived cysteine proteases and a major mite allergen, Der p 1 [23, 24]. These results indicated that SCCA molecules could play a role in a defense mechanism against extrinsic cysteine proteases downstream of the IL-4 or IL-13 signals. Furthermore, the IL-13 receptor α2 chain (IL-13Rα2), the component of the other type of IL-13R, was involved in the IL-4 or IL-13–inducible genes in bronchial epithelial cells in our microarray analysis [16]. Because we and other groups confirmed that IL-13Rα2 acts as a decoy receptor, inhibiting the IL-13 signals [25–27], these data indicated that there is a negative feedback regulation for the IL-13 signals in bronchial epithelial cells by induction of IL-13Rα2.

Conclusions

In summary, the microarray analyses that have been performed thus far have provided us not only with information on gene expression profiling, but also some novel pathogenic mechanisms of bronchial asthma. These studies teach us the importance of experimental design for microarray analysis and of the effort required to perform the functional analysis of the gene products. These strategies and experiences about the application of microarray technology in bronchial asthma should also be relevant to clarify the pathogenesis of various diseases other than bronchial asthma.

Acknowledgements
We thank Dr. Dovie R. Wylie for critical review of this manuscript. This work was supported by a grant-in-aid for Scientific Research from the Ministry of Education, Science, Sports and Culture of Japan.

References

1 Masoli M, Fabian D, Holt S, Beasley R (2004) The global burden of asthma: executive summary of the GINA Dissemination Committee report. *Allergy* 59: 469–478
2 Holgate ST (1999) The epidemic of allergy and asthma. *Nature* 402: B2–4
3 Wills-Karp M (2001) IL-12/IL-13 axis in allergic asthma. *J Allergy Clin Immunol* 107: 9–18
4 Izuhara K, Arima K, Yasunaga S (2002) IL-4 and IL-13: their pathological roles in allergic diseases and their potential in developing new therapies. *Curr Drug Targets Inflamm Allergy* 1: 263–269
5 Laprise C, Sladek R, Ponton A, Bernier MC, Hudson TJ, Laviolette M (2004) Functional classes of bronchial mucosa genes that are differentially expressed in asthma. *BMC Genomics* 5: 21
6 Lilly CM, Tateno H, Oguma T, Israel E, Sonna LA (2005) Effects of allergen challenge on airway epithelial cell gene expression. *Am J Respir Crit Care Med* 171: 579–586
7 Guajardo JR, Schleifer KW, Daines MO, Ruddy RM, Aronow BJ, Wills-Karp M, Hershey GK (2005) Altered gene expression profiles in nasal respiratory epithelium reflect stable *versus* acute childhood asthma. *J Allergy Clin Immunol* 115: 243–251
8 Zimmermann N, King NE, Laporte J, Yang M, Mishra A, Pope SM, Muntel EE, Witte

DP, Pegg AA, Foster PS et al (2003) Dissection of experimental asthma with DNA microarray analysis identifies arginase in asthma pathogenesis. *J Clin Invest* 111: 1863–1874
9 Karp CL, Grupe A, Schadt E, Ewart SL, Keane-Moore M, Cuomo PJ, Kohl J, Wahl L, Kuperman D, Germer S et al (2000) Identification of complement factor 5 as a susceptibility locus for experimental allergic asthma. *Nat Immunol* 1: 221–226
10 Munitz A, Bachelet I, Finkelman FD, Rothenberg ME, Levi-Schaffer F (2007) CD48 is critically involved in allergic eosinophilic airway inflammation. *Am J Respir Crit Care Med* 175: 911–918
11 Kuperman DA, Lewis CC, Woodruff PG, Rodriguez MW, Yang YH, Dolganov GM, Fahy JV, Erle DJ (2005) Dissecting asthma using focused transgenic modeling and functional genomics. *J Allergy Clin Immunol* 116: 305–311
12 Zou J, Young S, Zhu F, Gheyas F, Skeans S, Wan Y, Wang L, Ding W, Billah M, McClanahan T et al (2002) Microarray profile of differentially expressed genes in a monkey model of allergic asthma. *Genome Biol* 3: research0020
13 Zimmermann N, Rothenberg ME (2006) The arginine-arginase balance in asthma and lung inflammation. *Eur J Pharmacol* 533: 253–262
14 Lee JH, Kaminski N, Dolganov G, Grunig G, Koth L, Solomon C, Erle DJ, Sheppard D (2001) Interleukin-13 induces dramatically different transcriptional programs in three human airway cell types. *Am J Respir Cell Mol Biol* 25: 474–485
15 Chu EK, Cheng J, Foley JS, Mecham BH, Owen CA, Haley KJ, Mariani TJ, Kohane IS, Tschumperlin DJ, Drazen JM (2006) Induction of the plasminogen activator system by mechanical stimulation of human bronchial epithelial cells. *Am J Respir Cell Mol Biol* 35: 628–638
16 Yuyama N, Davies DE, Akaiwa M, Matsui K, Hamasaki Y, Suminami Y, Yoshida NL, Maeda M, Pandit A, Lordan JL et al (2002) Analysis of novel disease-related genes in bronchial asthma. *Cytokine* 19: 287–296
17 Cho SH, Tam SW, Demissie-Sanders S, Filler SA, Oh CK (2000) Production of plasminogen activator inhibitor-1 by human mast cells and its possible role in asthma. *J Immunol* 165: 3154–3161
18 Corry DB (1999) IL-13 in allergy: home at last. *Curr Opin Immunol* 11: 610–614
19 Wills-Karp M, Chiaramonte M (2003) Interleukin-13 in asthma. *Curr Opin Pulm Med* 9: 21–27
20 Izuhara K, Arima K, Kanaji S, Ohta S, Kanaji T (2006) IL-13: a promising therapeutic target for bronchial asthma. *Curr Med Chem* 13: 2291–2298
21 Takayama G, Arima K, Kanaji T, Toda S, Tanaka H, Shoji S, McKenzie AN, Nagai H, Hotokebuchi T, Izuhara K (2006) Periostin: a novel component of subepithelial fibrosis of bronchial asthma downstream of IL-4 and IL-13 signals. *J Allergy Clin Immunol* 118: 98–104
22 Silverman GA, Bird PI, Carrell RW, Church FC, Coughlin PB, Gettins PG, Irving JA, Lomas DA, Luke CJ, Moyer RW et al (2001) The serpins are an expanding superfamily

of structurally similar but functionally diverse proteins. Evolution, mechanism of inhibition, novel functions, and a revised nomenclature. *J Biol Chem* 276: 33293–33296

23 Kanaji S, Tanaka Y, Sakata Y, Takeshita K, Arima K, Ohta S, Hansell EJ, Caffrey C, Mottram JC, Lowther J et al (2007) Squamous cell carcinoma antigen 1 is an inhibitor of parasite-derived cysteine proteases. *FEBS Lett* 581: 4260–4264

24 Sakata Y, Arima K, Takai T, Sakurai W, Masumoto K, Yuyama N, Suminami Y, Kishi F, Yamashita T, Kato T et al (2004) The squamous cell carcinoma antigen 2 inhibits the cysteine proteinase activity of a major mite allergen, Der p 1. *J Biol Chem* 279: 5081–5087

25 Kawakami K, Taguchi J, Murata T, Puri RK (2001) The interleukin-13 receptor α2 chain: an essential component for binding and internalization but not for interleukin-13-induced signal transduction through the STAT6 pathway. *Blood* 97: 2673–2679

26 Chiaramonte mg, Mentink-Kane M, Jacobson BA, Cheever AW, Whitters MJ, Goad ME, Wong A, Collins M, Donaldson DD, Grusby MJ et al (2003) Regulation and function of the interleukin 13 receptor α2 during a T helper cell type 2–dominant immune response. *J Exp Med* 197: 687–701

27 Yasunaga S, Yuyama N, Arima K, Tanaka H, Toda S, Maeda M, Matsui K, Goda C, Yang Q, Sugita Y et al (2003) The negative-feedback regulation of the IL-13 signal by the IL-13 receptor α2 chain in bronchial epithelial cells. *Cytokine* 24: 293–303

Gene expression profiling in neurological and neuroinflammatory diseases

Sergio E. Baranzini

Department of Neurology, School of Medicine, University of California, San Francisco, Medical Sciences Building S-256, 513 Parnassus Ave., San Francisco, CA 94143, USA

Abstract

Neurological diseases arise when a sufficient number of neural cells cease to perform their normal functions, lose their ability to respond to the local environment, and die. In the last decade, a major technological leap led to the development of efficient and cost-effective high-throughput methods for determining gene expression. This in turn resulted in rapid accumulation of data describing gene expression patterns in human and experimental animal brains. In this chapter, I review several of the latest advances in large-scale gene expression in neurological and neuroinflammatory disorders, including Alzheimer's disease, Parkinson's disease, multiple sclerosis, and epilepsy. The need for integration of different sources of data is discussed in the context of systems biology, and I also address how such integration could result in improved diagnostics, therapies, and disease prevention.

Introduction

Advancement and evolution in biomedical research are supported not only through systematic accumulation of knowledge but also through development of new technologies. Although traditional methods for assaying gene expression enabled researchers to interrogate a relatively small number of genes at a time, new tools have emerged in the last decade that allow investigators to address previously intractable problems and to uncover novel potential targets for therapies. Today, microarrays allow an experimenter to interrogate the expression of thousands of genes from several biological samples quickly and efficiently. DNA microarrays are a major methodological advance and illustrate how new technologies provide powerful tools for researchers. Since the original description of the method in 1995, the number of experiments using DNA microarrays has grown exponentially, and their development can be compared to that of the polymerase chain reaction in the early 1980s. Scientists are currently using microarray-based technologies to try to understand fundamental aspects of growth and development, as well as to explore the underlying genetic causes of many human diseases. More recent applications

Microarrays in Inflammation, edited by Andreas Bosio and Bernhard Gerstmayer

include testing drug toxicity in the pharmaceutical industry, monitoring pharmacological interventions, and predicting therapeutic outcomes.

Cells grow and differentiate in the context of homeostatic balance, a condition required for organism survival. Within variable margins of tolerance according to spatial and temporal coordinates dictated by developmental processes, a given cell will express a particular set of genes that can be viewed as a proxy for its function. Diseases occur when the molecular and cellular mechanisms in charge of maintaining homeostasis fail. Such misbalance may be reflected by an altered pattern of gene expression. Therefore, gene expression profiling can be performed to determine the regulatory state of a cell or tissue. In neurological diseases, identification of such patterns is hampered by the complex anatomy and broad functionality of the brain, in which the affected area can be difficult to access or may not be known. However, significant advances have been made through the use of technologies, permitting inspection of only affected areas, as well as through the use of experimental animals. In the sections that follow, I discuss examples of how transcriptional profiling may help elucidate the pathogenesis of neurodegenerative diseases, psychiatric disorders, and traumatic brain injury.

Alzheimer's disease

Alzheimer's disease (AD) is the most common cause of dementia and affects ~5% of people older than 60 years of age worldwide [1]. Patients with AD present with subtle changes in behavior and memory loss, and the symptoms gradually increase in severity, resulting in total loss of cognitive functions. Histological characterization of AD is usually performed postmortem; in AD, senile plaques and neurofibrillary tangles (NFTs) are present throughout the brain. Several animal models based on transgenic technology are available that reproduce various aspects of the disease. These mice show some of the characteristic AD pathology, such as age-dependent formation of amyloid plaques consisting of amyloid-β peptides. However, the mice usually lack both the tau pathology (i.e., NFT formation) and the neurodegeneration associated with AD [2].

In both humans and experimental animals, the area of the brain most commonly used for molecular studies in AD is the hippocampus [3–5]. Initial reports of human studies describe the use of at least a few milligrams of tissue to obtain high yields of RNA for labeling and hybridization. Because of the different tissues and microarray platforms used, a direct comparison among these results is difficult. However, it was observed that, compared with tissues from controls, tissues from AD patients generally contained a higher proportion of inflammation- and apoptosis-related transcripts and a lower proportion of signal transduction and gene transcription [4, 6, 7]. In another study, changes in 89 genes were shown to correlate with both premortem mental status and the NFT index [7]. Changes in transcription factors and signaling genes regulating proliferation and differentiation, oligodendrocyte growth factors, and protein kinase A pathway genes seemed to be part of a coregulated transcriptional network. On the basis of these findings, the investigators proposed an intriguing hypothesis for AD pathogenesis, in which early axonopathy is induced by a tumor suppressor-like mechanism and oligodendrocyte stimulation, resulting in spreading of the disease process along myelinated nerve fibers. This hypothesis seems to be confirmed by a recent study showing that hippocampi from patients with moderate dementia display more inflammatory changes than those from AD patients without dementia [8].

Given that not all cells of a particular brain region are equally affected, much larger gene expression differences can be detected when homogeneous samples or even single cells are studied. In a recent report, individual CA1 neurons harboring NFTs showed a significant reduction in several classes of messenger RNAs known to encode proteins implicated in AD neuropathology, including protein phosphatases/kinases, cytoskeletal elements, synaptic-related markers, glutamate receptors, and dopamine receptors [9]. The amyloid-β precursor protein (APP) gene was found to be downregulated about two-fold. In contrast, the acid hydrolase cathepsin D was the most upregulated gene in CA1 neurons. These findings are consistent with a growing body of literature indicating that activation of the lysosomal protease system is an early feature in AD and may prove to be an important biomarker of the disease [10–12].

Although the hippocampus seems to be preferentially affected in early AD, similar pathological gene expression patterns have also been identified in inferior parietal lobes [13] and in the thalamus and primary visual cortex in advanced cases [14], suggesting a spreading of the abnormal expression signature throughout the brain. The finding can be generalized to other neurodegenerative disorders, such as multiple sclerosis (MS) and Parkinson's disease (PD).

In a recent study performed in blood-derived mononuclear cells, AD patients exhibited a significant decline in the expression of genes concerned with cytoskeletal maintenance, cellular trafficking, cellular stress response, redox homeostasis, transcription, and DNA repair. Decreased expression of several genes that may affect amyloid-β production and the processing of the microtubule-associated protein tau was also observed [15].

Parkinson's disease

PD is a neurodegenerative disease that primarily affects dopaminergic regions of the substantia nigra. Inflammation is a common finding in the PD brain, but because of the limitation of postmortem analyses, its relationship to disease progression has been difficult to establish. Although no spontaneous model exists for PD, injection of specific neurotoxins (6-OHDA or MPTP) in experimental animals accurately reproduces several clinical and histopathological aspects of the human disease [16–18]. Gene expression analysis of postmortem substantia nigra identified chaperones, ubiquitination, vesicle trafficking, and nuclear-encoded mitochondrial genes as the main entities and functions affected in PD patients [19]. This finding correlates well with results of previous analyses performed in animal models.

In another study in substantia nigra, decreased expression of 69 genes and upregulation of 68 genes were detected [20]. Classification into functional groups revealed that the downregulated genes were related to signal transduction, protein degradation (e.g., ubiquitin-proteasome subunits), dopaminergic transmission/metabolism, iron transport, protein modification/phosphorylation, and energy pathways. The decreased expression of five subunits of the ubiquitin-proteasome system (UPS), SKP1A, a member of the SCF (E3) ubiquitin-ligase complex, and chaperone HSC-70 was of interest, because these can lead to extensive impairment in the function of an entire repertoire of proteins. The upregulated genes were identified as members of cell adhesion/cytoskeleton, extracellular matrix components, cell cycle, protein modification/phosphorylation, protein metabolism and transcription, and inflammation/hypoxia pathways. These findings were confirmed and expanded in a recent study, in which an overall decrease in expression affecting the majority of mitochondrial and UPS genes was detected [21]. The downregulated transcripts identified in this study also included genes encoding subunits of mitochondrial complex I and the PD-linked UCHL1. It was hypothesized that the mitochondria and the proteasome form a higher-order gene regulatory network that is severely perturbed in PD.

When total striatum of treated mice was used as a source of tissue for molecular studies, changes in signaling cascade proteins, transport proteins, and transcription factors were observed, regardless of the chemical used for inducing the disease. However, when expression changes caused by different induction methods were compared, marked differences in the expression of several genes was noted, suggesting a different mechanism of action for each of these toxins [22].

Most microarray-based studies in PD use either rodent tissue or cultured cells as starting material. The reason for this may be that suitable human tissue is lacking, because most neurons affected by PD die. With the advancement of technologies aimed at single-cell isolation and RNA amplification, it is likely that molecular profiling of PD-affected human tissue will soon be a reality.

Multiple sclerosis

MS is a common inflammatory disorder of the central nervous system (CNS) that is characterized by myelin loss, gliosis, varying degrees of axonal pathology, and progressive neurological dysfunction. A large body of research supports a multifactorial etiology, with an underlying genetic susceptibility likely acting in concert with undefined environmental exposures [23–25]. Although the exact pathogenic mechanisms underlying MS remains unknown, it has been proposed that lymphocytes activated in the periphery by a microbial-like antigen migrate to the CNS, become attached to receptors on endothelial cells, and then cross the blood–brain barrier (BBB) directly into the interstitial matrix. T cells are then reactivated *in situ* by fragments of myelin antigens exposed in the context of major histocompatibility complex molecules on the surface of antigen-presenting cells (macrophages, microglia, and perhaps astrocytes). Reactivation induces the release of proinflammatory cytokines that open the BBB further and stimulate chemotaxis, resulting in a second, larger wave of inflammatory cell recruitment and leakage of pathogenic antibodies and other plasma proteins into the nervous system.

In one of the first attempts to analyze the transcriptional profile of the MS lesion, researchers used a normalized cDNA library from a brain sample obtained postmortem from a patient with the primary progressive form of the disease [26]. This study, albeit not statistically powerful, identified several inflammatory genes and known putative autoantigens in the MS-derived library but not in two normal control libraries. In a follow-up report, the authors described the differential expression of 62 genes, including the Duffy chemokine receptor, interferon regulatory factor 2, and tumor necrosis factor alpha [27]. The absence of total or even partial overlapping in the genes identified by these two screens is noteworthy. This observation highlights the wide variability usually found in large-scale gene expression profiling when different experimental platforms are employed.

Using a combination of large-scale expressed sequence tag (EST) sequencing and microarray technology, we noted differentially expressed genes in active lesions and control brains from MS patients and experimental animals [28]. In that study, alpha B-crystallin was identified as the most abundant transcript unique to MS plaques. Alpha B-crystallin is an inducible heat shock protein localized in the myelin sheath, and a putative target for T cells in MS [29]. The next five most abundant transcripts included those for prostaglandin D synthase, prostatic binding protein, ribosomal protein L17, and osteopontin (OPN). When studied in brain lesions from the animal model experimental autoimmune encephalomyelitis (EAE), OPN was found to be expressed broadly in microglia near perivascular inflammatory lesions during both relapses and remission from disease. In addition, OPN-deficient mice were resistant to progressive EAE and had frequent remissions. Finally, myelin-reactive T cells from $OPN^{-/-}$ mice produced more interleukin (IL)-10 and less interferon-γ and IL-12 than those from $OPN^{+/+}$ mice. Taken together, our studies suggest that OPN may regulate Th1 responses involved in CNS autoimmunity and may be an attractive target for new therapies designed to block development of progressive MS.

In another study, brain samples from 15 patients with relapsing-remitting MS and 15 controls were analyzed by probing cDNA arrays bound to a nylon membrane with radiolabeled cDNA from each specimen [30]. Only 34 of 4000 genes interrogated showed statistically significant differential expression in MS samples compared with controls. Surprisingly, only a very small fraction of these genes can be directly associated with current models of MS pathogenesis or with its downstream inflammatory effects. It is thus conceivable that cross-sectional experimental approaches often prevent a clear sorting out of bona fide regulatory changes in gene expression because of the low signal-to-noise ratio characteristic of molecular control processes.

Several microarray-based studies aimed at describing the molecular signature of the MS lesion have been recently published [31–35]. Although difficult to compare owing to differences in plaque activity, tissue harvesting, processing, and analysis, a common picture seems to emerge from these studies. On the one hand, a clear upregulation of inflammation-related genes (mainly HLA class II and immunoglobulin genes) can be observed, particularly at the edge of active plaques. This fact likely reflects an active immune response in the lesions. On the other hand, these analyses reveal upregulation of genes involved in maintenance of cellular homeostasis, and in neural protective mechanisms, a response similar to that observed after ischemic insult. As in other neuroinflammatory diseases, activation of these common pathways seems to be a general response by damaged tissue to restore homeostasis.

The expression of immunoglobulin genes showed differential regulation according to the stage of the plaque (i.e., a higher expression was observed in acute than in chronic silent lesions). Similarly, granulocyte colony-stimulating factor (G-CSF) was upregulated in most of the active plaque types. This finding was validated *in vivo* using the EAE model. In that experiment, animals that had previously received a subcutaneous dose of G-CSF developed a much milder disease than did control animals, suggesting an active immunomodulatory mechanism upregulating the production of this trophic factor in EAE and MS lesions [35]. Genes with decreased expression included those for several myelin components, such as proteolipid protein (PLP), myelin-associated glycoprotein, and myelin oligodendrocyte glycoprotein (MOG). This last finding may reflect not only the catabolic demyelinating process but also ineffective or absent myelin repair. One obvious limitation of these studies, as in several other brain disorders discussed earlier, is the analysis of postmortem tissue, which provides an "instant picture" of a much longer, dynamic process. Another important caveat concerns the heterogeneity of tissue under study. In studies that focus on particular areas of the brain, the anatomical complexity of tissue analyzed is reduced.

Several studies have used the EAE model to better characterize the molecular signature of the neuroinflammatory lesion [36–39]. More homogeneous results were obtained across different studies using the EAE model than across those performed on MS plaques. This is not surprising, given the flexibility offered by animal models in terms of controlling variables such as time of tissue harvesting, genetic homogeneity, and reproducibility in tissue harvesting. The general pattern observed in these studies is that inflammation-related genes are dramatically upregulated rather early in the disease, and their expression wanes at later stages. Concurrently, most CNS-specific genes differentially expressed appear to be downregulated, potentially reflecting a neural response to the inflammatory insult [38–40]. These findings were recently extended to humans in a study that investigated gene expression in the motor cortex of MS subjects [41].

Several studies have also investigated the gene expression signature in peripheral blood cells from MS patients. These studies sought to explore to what extent immune cells carry information about ongoing inflammatory processes in the brain. Surprisingly, and in contrast to what was observed in studies of neural tissue, downregulated genes in T cells generally outnumbered upregulated genes. Most notably, an altered regulation of apoptosis-related transcripts and downregulation of transcripts involved in DNA repair was observed in these cells, suggesting an impaired balance of critical functions for cell survival [42]. In addition to providing a more convenient source of RNA than neural tissue, gene expression profiling from blood also permits longitudinal analysis, classification of disease subtypes, and prediction of disease progression. These advantages of studying blood material are highlighted by a recent study in which MS patients were classified according to their T cell gene expression into several groups that correlated with clinical subtypes and response to immunomodulatory therapy [43].

Schizophrenia

Schizophrenia is a chronic psychiatric illness, affecting nearly 0.4% of the general population [1]. Unlike in AD, PD, or MS, there is no morphologically defined lesion or other neuropathological hallmark in schizophrenia, a deficiency that complicates efforts to identify the affected brain region. Molecular studies have identified changes in presynaptic release [43, 44], metabolic pathways [45], γ-aminobutyric acid (GABA)-glutamate transcripts [46], and altered oligodendrocyte function [47]. One gene identified in these studies encodes the regulator of G-protein signaling 4 (RGS4). This gene belongs to a family of regulatory molecules that act as GTPase-activating proteins for G alpha subunits of heterotrimeric G proteins. Subsequent genetic association studies showed a distortion of transmission for several RGS4 haplotypes, suggesting that RGS4 is indeed a schizophrenia susceptibility gene [48–50]. It was recently suggested that the defect in myelination found in brains from schizophrenic individuals may be due to altered oligodendrocyte development [51]. Along these lines, a recent study found a change in expression of proteolipid protein 1 (PLP1) across several individuals with schizophrenia [52]. Given these converging findings, it was proposed that decreased PLP1 expression may be a significant contributor to decreased expression of oligodendrocyte development in the prefrontal cortex in schizophrenia [52].

Memory and learning

As in schizophrenia, there are no major morphological changes during memory formation that can be easily identified in the mammalian brain. Nevertheless, extensive stimulation and recording data have implicated the cortex of the cerebellum and the hippocampus as areas where signals involved in learning and memory formation are processed [53–55]. In a microarray experiment performed in rabbits, very few changes were observed when expression data from paired animals were compared [56]. Interestingly, a large proportion of the differentially expressed genes were downregulated. The investigators speculated that such changes may be an end point to a dynamic gene expression process recruited during the acquisition and retention of memory. Alternatively, memory storage may require reduced expression of proteins that exert inhibitory constraints.

In another study, microarrays were used to examine differences in hippocampal gene expression between two F1 hybrid mouse strains that perform well on the Morris water maze and two inbred strains that perform poorly [57]. Few of the genes discovered in this analysis (*Wfs1*, *Nub1*, *Rab2a*, *Prpsap2*, and *Slc9a3r2*) have known functions, and these genes may play a role in hippocampal learning and memory.

Seizure disorders

Seizure disorders encompass a spectrum of neurological abnormalities, including acute seizures triggered by a number of circumstances (e.g., drugs, electrolyte disturbances, infections) and recurrent, spontaneous seizures that characterize *epilepsy*. As in PD, a spontaneous experimental model of epilepsy is not available, and therefore investigators use specific chemicals such as kainic acid (a stimulant of a subset of glutamate receptors) and pilocarpine (an acetylcholine receptor agonist) to induce the disease. These models reproduce at least some aspects of one of the most prevalent human seizure disorders, temporal lobe epilepsy (TLE). One of the most fundamental levels of seizure-associated alteration in brain function is that of gene expression. Early studies of gene regulation in sensitized animals highlighted the influence of neuropeptides and proenkephalin expression a few hours after the onset of seizures [58–60]. In contrast, expression of the immediate-early gene c-*fos* was detected 15 minutes after induction of seizures, indicating a remarkably fast cellular response to stimulation [61]. Given the participation of c-*fos* in the AP-1 transcription factor complex, this increased expression of c-*fos* raised the possibility that this is an important first step in the seizure-induced regulation of other genes. This idea was supported by the finding that AP-1 can regulate expression of proenkephalin [62]. Recent microarray-based gene expression analyses have identified several genes whose expression changes significantly after induced seizures [63–65].

In general, a predominance of injury response and cell survival genes was observed, consistent with the hypothesis that neuronal regeneration and network reorganization are needed after an epileptic seizure. When gene expression in specific areas of the brain in normal young animals was compared with that in epileptic animals, a common pattern was observed, mostly dominated by metabolism-related genes [65]. These findings are consistent with the hypothesis that expression programs geared toward high metabolic demands characteristic of neural development are activated shortly after an epileptic seizure.

Patients with TLE often have a shrunken hippocampus, which is known to be where seizures originate. Sclerotic hippocampi showed increased expression of several transcripts associated with molecular signaling pathways, including astrocyte structure (GFAP, NF2, PALLD), calcium regulation (S100B, CXCR4) and BBB function (AQP4, CCR2, CCL3, PLEC1) and inflammation [66]. The authors hypothesized that astrocytes in sclerotic tissue have activated molecular pathways that could lead to enhanced release of glutamate by these cells. Such glutamate release may excite surrounding neurons and elicit seizure activity. In a study using entorhinal cortex of individuals with mesial temporal lobe epilepsy, consistent downregulation of neuropeptide Y receptor (NPY) was observed [67]. Given that NPY is the most abundant neurotransmitter in the brain, this finding suggests that impaired synaptic transmission is a characteristic feature of this disorder.

Gene expression profiling of biopsied temporal lobe cortices from TLE patients showing active signs of the disease (electrophysiological spikes) revealed a downregulation of multiple GABAergic genes (*GABRA5*, *GABRB3*, *ABAT*) and an upregulation of oligodendrocyte and lipid metabolism transcripts (MOG, CA2, CNP, SCD, PLP1, FA2H, ABCA2). In addition, several transcripts related to the classical MAPK cascade showed expression-level alterations between spiking and nonspiking samples (G3BP2, MAPK1, PRKAR1A, and MAP4K4) [68]. The authors observed that abnormal electrical brain activity in the spiking samples was strongly correlated with gene expression changes, and they speculated that some of the observed transcriptional changes may be directly involved in the induction or prevention of ictal events in epilepsy.

In humans, brain trauma is a risk factor for epilepsy. In a *traumatic brain injury* (TBI) study using an experimental model of brain damage, in which a controlled physical force was used to provoke concussion, an almost instantaneous expression of an immediate-early gene followed by metabolism-related genes was observed. This result mimics findings noted in animals with chemically induced seizures. In addition, a large wave of inflammation-related genes is characteristic of TBI, possibly as a result of BBB disruption [69–71]. Also, genes coding for neurotrophic factors (NTFs) such as brain-derived neurotrophic factor, nerve growth factor, and neurotrophin-3 were found to play an important role in regeneration of the injured tissue. NTFs provide trophic support during development and adult life, promote neuronal survival, and restore neuronal connections by promoting axonal outgrowth. These findings underscore the complexity and heterogeneity of pathological and molecular responses to TBI. Clearly, a better understanding of the molecular pathways involved in neural regeneration and plasticity is needed to develop better treatment strategies.

Ischemic stroke

Stroke is a major cause of death and disability in the world. Atherosclerotic plaques of various sizes and depositions are commonly seen in the adult population, and rupture of a carotid artery plaque followed by embolization is a major cause of stroke. Very little is known about the factors that predispose atherosclerotic plaques to instability, rupture, and embolization. Diagnosis and management of stroke must be improved: up to 30% of initial diagnoses of this condition may be incorrect, and such misdiagnoses entail the costs of added investigations and management. Knowledge obtained through gene expression profiling could serve as additional and personalized biomarker information.

An important factor to consider in applying gene expression methodology to stroke and other neurological disorders is that brain tissue is rarely available or accessible for study. Therefore, investigators have used peripheral blood as a most practical and widely available source of RNA. The rationale for using peripheral blood is that after stroke, there is a well-characterized inflammatory response consisting of the selective migration and infiltration of blood-borne white blood cells into the brain, involving all white blood cell types.

The first study of gene expression in humans was carried out in peripheral blood mononuclear cells from 20 individuals shortly after a stroke event [72]. The expression of 190 genes was found to be significantly different between the stroke and control groups. Of these 190 genes, the authors identified genes involved in white blood cell activation and differentiation (approximately 60%), genes associated with hypoxia and vascular repair, and genes potentially associated with an altered cerebral microenvironment. When the list of differentially expressed genes in stroke was compared with a list of genes detected in the blood of individuals with other inflammatory disorders (sickle cell disease and MS), none of the genes overlapped. This could suggest that each inflammatory disease displays a characteristic signature. However, given the heterogeneity in sample collection and preparation, microarray platforms, and analyses, the absence of commonalities among these studies is not entirely surprising. A more recent study, using blood samples obtained at several times after stroke, revealed an increase in the expression of MMP9, S100A12, coagulation factor V, arginase I, carbonic anhydrase IV, and CD96 within hours of the ischemic attack [73]. The authors concluded that although the mechanisms for the extremely rapid changes in gene expression after ischemic stroke remain to be elucidated, they may involve direct interactions between the white blood cells and ischemic brain endothelium.

Although gene expression profiling has the potential to provide useful information for diagnosis and management of acute ischemic stroke, the specificity of gene signatures associated with this condition needs to be further determined. The potential clinical uses of this new methodology include prediction of outcome and determination of hemorrhagic risk after thrombolytic therapy with recombinant tissue plasminogen activator.

Systems biology

By studying expression patterns in pathological tissue or in selected areas of the brain, we have acquired a large body of information about the most salient features of each system (e.g., inflammation in MS, neuronal death in PD, neural plasticity in epilepsy). However, the link between this information and effective translational knowledge is still missing.

Systems biology is a holistic approach to understanding biological systems. It proposes integration of knowledge across various disciplines to elucidate fundamental design principles that have allowed living organisms to evolve highly complex structures and functions [74, 75]. The ultimate goal of this system-based approach is the generation of *in silico* disease models that can accurately reproduce known experimental results but also have predictive power to permit the development of more effective therapies.

Toward personalized medicine

A majority of therapeutic decisions today are based on an incomplete characterization of the patient and the disease, such that the therapy most likely to succeed is often not known and therefore not prescribed. Improvements through more in-depth patient characterization and individually targeted therapeutic decision-making should result in improved patient well-being and reduced costs of long-term medical care [76]. Drugs and therapies that work very well in only specific forms of a disease may be more effectively targeted when there is more readily accessible knowledge about which individuals these drugs are suitable for. Moreover, toxic side effects in small subpopulations have blocked the application of otherwise effective treatments because individuals showing the adverse effects were not readily identified and characterized.

There is an expanding array of databases covering clinical patient histories, new treatment trial data, and detailed measurements on patient clinical assays, genetic variation, and genomic molecular activity profiling. The information in these databases is allowing specialists to define patient profiles in increasing detail, linking microscopic with macroscopic patient variables. By introducing such a systems-level characterization, analysts are beginning to capture diagnostic, prognostic, and causal variables simultaneously as one set. Availability of this information is invaluable for successful downstream data mining and predictive modeling.

The ability to apply patient data to the learning process through models will allow us to understand new processes and apply this new knowledge at the bedside.

Concluding remarks

The last few years have seen an impressive advancement in the accumulation of biological information in almost all aspects of neurological disorders. Although each disease is likely to have its own pathogenic mechanism, it is not uncommon to find that many of the genes and pathways discovered to be altered in one neurological disease are also deregulated in another, not necessarily sharing the same pathophysiology. Furthermore, although several immunological responses and mechanisms of tissue damage repair take place at different times during the course of one neurodegenerative disease, these pathways are likely to be different from the pathways in charge of maintaining homeostasis in the healthy brain. It is therefore plausible that physiological systems (including the CNS) have a restricted ability to deal with genetic or environmental perturbations, and thus different insults (e.g., trauma, ischemic, metabolic) trigger a limited number of pathways, such as inflammation, oxidative stress, apoptosis, protein folding and degradation impairment, or energy starvation, ultimately leading to common phenotypes (e.g., parkinsonism, dementia, seizures).

The challenge before us is to integrate all the accumulated sources of information and translate them into effective knowledge, in order to develop better therapies that will have a significant impact on the quality of life of neurological patients.

References

1 Prentice T (2001) *The World Health Report 2001. Mental health: new understanding, new hope.* World Health Organization, Geneva

2 Higgins GA, Jacobsen H (2003) Transgenic mouse models of Alzheimer's disease: phenotype and application. *Behav Pharmacol* 14, 419–438

3 Hata R et al (2001) Up-regulation of calcineurin Abeta mRNA in the Alzheimer's disease brain: assessment by cDNA microarray. *Biochem Biophys Res Commun* 284, 310–316

4 Colangelo V et al (2002) Gene expression profiling of 12633 genes in Alzheimer hippocampal CA1: transcription and neurotrophic factor down-regulation and up-regulation of apoptotic and pro-inflammatory signaling. *J Neurosci* Res 70, 462–473

5 Yao PJ et al (2003) Defects in expression of genes related to synaptic vesicle trafficking in frontal cortex of Alzheimer's disease. *Neurobiol Dis* 12, 97–109
6 Reddy PH et al (2004) Gene expression profiles of transcripts in amyloid precursor protein transgenic mice: up-regulation of mitochondrial metabolism and apoptotic genes is an early cellular change in Alzheimer's disease. *Hum Mol Genet* 13, 1225–1240
7 Blalock EM et al (2004) Incipient Alzheimer's disease: microarray correlation analyses reveal major transcriptional and tumor suppressor responses. *Proc Natl Acad Sci USA* 101, 2173–2178
8 Parachikova A et al (2006) Inflammatory changes parallel the early stages of Alzheimer disease. *Neurobiol Aging* 28, 1821–1833
9 Ginsberg SD et al (2000) Expression profile of transcripts in Alzheimer's disease tangle-bearing CA1 neurons. *Ann Neurol* 48, 77–87
10 Cataldo AM et al (1995) Gene expression and cellular content of cathepsin D in Alzheimer's disease brain: evidence for early up-regulation of the endosomal-lysosomal system. *Neuron* 14, 671–680
11 Cataldo AM et al (1996) Properties of the endosomal-lysosomal system in the human central nervous system: disturbances mark most neurons in populations at risk to degenerate in Alzheimer's disease. *J Neurosci* 16, 186–199
12 Nixon RA et al (2000) The endosomal-lysosomal system of neurons in Alzheimer's disease pathogenesis: a review. *Neurochem Res* 25, 1161–1172
13 Weeraratna AT et al (2007) Alterations in immunological and neurological gene expression patterns in Alzheimer's disease tissues. *Exp Cell Res* 313, 450–461
14 Cui JG et al (2007) Expression of inflammatory genes in the primary visual cortex of late-stage Alzheimer's disease. *Neuroreport* 18, 115–119
15 Maes OC et al (2006) Transcriptional profiling of Alzheimer blood mononuclear cells by microarray. *Neurobiol Aging* 28, 1795–1809
16 Perese DA et al (1989) A 6–hydroxydopamine-induced selective parkinsonian rat model. *Brain Res* 494, 285–293
17 Grunblatt E et al (2001) Gene expression analysis in N-methyl-4-phenyl-1,2,3,6-tetrahydropyridine mice model of Parkinson's disease using cDNA microarray: effect of R-apomorphine. *J Neurochem* 78, 1–12
18 Mandel S et al (2003) Genes and oxidative stress in parkinsonism: cDNA microarray studies. *Adv Neurol* 91, 123–132
19 Hauser MA et al (2005) Expression profiling of substantia nigra in Parkinson disease, progressive supranuclear palsy, and frontotemporal dementia with parkinsonism. *Arch Neurol* 62, 917–921
20 Mandel S et al (2005) Gene expression profiling of sporadic Parkinson's disease substantia nigra pars compacta reveals impairment of ubiquitin-proteasome subunits, SKP1A, aldehyde dehydrogenase, and chaperone HSC-70. *Ann NY Acad Sci* 1053, 356–375
21 Duke DC et al (2006) Transcriptome analysis reveals link between proteasomal and mitochondrial pathways in Parkinson's disease. *Neurogenetics* 7, 139–148

22 Holtz WA, O'Malley KL (2003) Parkinsonian mimetics induce aspects of unfolded protein response in death of dopaminergic neurons. *J Biol Chem* 278, 19367–19377
23 Hauser SL, Goodin DS (2005) Multiple sclerosis and other demyelinating diseases. In: Braunwald et al (eds): *Harrison's Principles in Internal Medicine.* McGraw-Hill, New York, 2461–2471
24 Compston A (1999) The genetic epidemiology of multiple sclerosis. *Philos Trans R Soc Lond B Biol Sci* 354, 1623–1634.
25 Oksenberg JR et al (2001) Multiple sclerosis: genomic rewards. *J Neuroimmunol* 113, 171–184
26 Becker KG et al (1997) Analysis of a sequenced cDNA library from multiple sclerosis lesions. *J Neuroimmunol* 77, 27–38
27 Whitney LW et al (1999) Analysis of gene expression in mutiple sclerosis lesions using cDNA microarrays. *Ann Neurol* 46, 425–428
28 Chabas D et al (2001) The influence of the proinflammatory cytokine, osteopontin, on autoimmune demyelinating disease. *Science* 294, 1731–1735
29 van Noort JM et al (1995) The small heat-shock protein alpha B-crystallin as candidate autoantigen in multiple sclerosis. *Nature* 375, 798–801
30 Ramanathan M et al (2001) *In vivo* gene expression revealed by cDNA arrays: the pattern in relapsing-remitting multiple sclerosis patients compared with normal subjects. *J Neuroimmunol* 116, 213–219.
31 Mycko MP et al (2003) cDNA microarray analysis in multiple sclerosis lesions: detection of genes associated with disease activity. *Brain* 126, 1048–1057
32 Graumann U et al (2003) Molecular changes in normal appearing white matter in multiple sclerosis are characteristic of neuroprotective mechanisms against hypoxic insult. *Brain Pathol* 13, 554–573
33 Mycko MP et al (2004) Microarray gene expression profiling of chronic active and inactive lesions in multiple sclerosis. *Clin Neurol Neurosurg* 106, 223–229
34 Lindberg RL et al (2004) Multiple sclerosis as a generalized CNS disease-comparative microarray analysis of normal appearing white matter and lesions in secondary progressive MS. *J Neuroimmunol* 152, 154–167
35 Lock C et al (2002) Gene-microarray analysis of multiple sclerosis lesions yields new targets validated in autoimmune encephalomyelitis. *Nat Med* 8, 500–508.
36 Matejuk A et al (2003) CNS gene expression pattern associated with spontaneous experimental autoimmune encephalomyelitis. *J Neurosci* Res 73, 667–678
37 Whitney LW et al (2001) Microarray analysis of gene expression in multiple sclerosis and EAE identifies 5–lipoxygenase as a component of inflammatory lesions. *J Neuroimmunol* 121, 40–48
38 Carmody RJ et al (2002) Genomic scale profiling of autoimmune inflammation in the central nervous system: the nervous response to inflammation. *J Neuroimmunol* 133, 95–107
39 Mix E et al (2004) Gene-expression profiling of the early stages of MOG-induced EAE proves EAE-resistance as an active process. *J Neuroimmunol* 151, 158–170

40 Baranzini SE et al (2005) Modular transcriptional activity characterizes the initiation and progression of autoimmune encephalomyelitis. *J Immunol* 174, 7412–7422
41 Dutta R et al (2006) Mitochondrial dysfunction as a cause of axonal degeneration in multiple sclerosis patients. *Ann Neurol* 59, 478–489
42 Satoh J et al (2005) Microarray analysis identifies an aberrant expression of apoptosis and DNA damage-regulatory genes in multiple sclerosis. *Neurobiol Dis* 18, 537–550
43 Mirnics K et al (2000) Molecular characterization of schizophrenia viewed by microarray analysis of gene expression in prefrontal cortex. *Neuron* 28, 53–67
44 Hemby SE et al (2002) Gene expression profile for schizophrenia: discrete neuron transcription patterns in the entorhinal cortex. *Arch Gen Psychiat*ry 59, 631–640
45 Middleton FA et al (2002) Gene expression profiling reveals alterations of specific metabolic pathways in schizophrenia. *J Neurosci* 22, 2718–2729
46 Vawter MP et al (2002) Microarray analysis of gene expression in the prefrontal cortex in schizophrenia: a preliminary study. *Schizophr Res* 58, 11–20
47 Tkachev D et al (2003) Oligodendrocyte dysfunction in schizophrenia and bipolar disorder. *Lancet* 362, 798–805
48 Chowdari KV et al (2002) Association and linkage analyses of RGS4 polymorphisms in schizophrenia. *Hum Mol Genet* 11, 1373–1380
49 Williams NM et al (2004) Support for RGS4 as a susceptibility gene for schizophrenia. *Biol Psychiatry* 55, 192–195
50 Morris DW et al (2004) Confirming RGS4 as a susceptibility gene for schizophrenia. *Am J Med Genet* 125B, 50–53
51 Hakak Y et al (2001) Genome-wide expression analysis reveals dysregulation of myelination-related genes in chronic schizophrenia. *Proc Natl Acad Sci USA* 98, 4746–4751
52 Pongrac J et al (2002) Gene expression profiling with DNA microarrays: advancing our understanding of psychiatric disorders. *Neurochem Res* 27, 1049–1063
53 Yeo CH et al (1985) Classical conditioning of the nictitating membrane response of the rabbit. I. Lesions of the cerebellar nuclei. *Exp Brain Res* 60, 87–98
54 Sweeney JA et al (1996) Positron emission tomography study of voluntary saccadic eye movements and spatial working memory. *J Neurophysiol* 75, 454–468
55 Ohta M (1998) [Changes in spatio-temporal patterns after long-term potentiation (LTP) in mouse hippocampal slices and effects of trichloroethylene on LTP]. *Hokkaido Igaku Zasshi* 73, 365–378
56 Cavallaro S et al (2001) Gene expression profiles during long-term memory consolidation. *Eur J Neurosci* 13, 1809–1815
57 Leil TA et al (2002) Finding new candidate genes for learning and memory. *J Neurosci* Res 68, 127–137
58 Bajorek JG et al (1986) Neuropeptides: anticonvulsant and convulsant mechanisms in epileptic model systems and in humans. *Adv Neurol* 44, 489–500
59 White JD, Gall CM (1986) Increased enkephalin gene expression in the hippocampus following seizures. *NIDA Res Monogr* 75, 393–396
60 White JD, Gall CM (1987) Differential regulation of neuropeptide and proto-onco-

gene mRNA content in the hippocampus following recurrent seizures. *Brain Res* 427, 21–29
61 Morgan JI et al (1987) Mapping patterns of c-fos expression in the central nervous system after seizure. *Science* 237, 192–197
62 Sonnenberg JL et al (1989) Regulation of proenkephalin by Fos and Jun. *Science* 246, 1622–1625
63 French PJ et al (2001) Seizure-induced gene expression in area CA1 of the mouse hippocampus. *Eur J Neurosci* 14, 2037–2041
64 Yun J et al (2003) Gene expression profile of neurodegeneration induced by alpha1B-adrenergic receptor overactivity: NMDA/GABAA dysregulation and apoptosis. *Brain* 126, 2667–2681
65 Elliott RC et al (2003) Overlapping microarray profiles of dentate gyrus gene expression during development- and epilepsy-associated neurogenesis and axon outgrowth. *J Neurosci* 23, 2218–2227
66 Lee TS et al (2007) Gene expression in temporal lobe epilepsy is consistent with increased release of glutamate by astrocytes. *Mol Med* 13, 1–13
67 Jamali S et al (2006) Large-scale expression study of human mesial temporal lobe epilepsy: evidence for dysregulation of the neurotransmission and complement systems in the entorhinal cortex. *Brain* 129, 625–641
68 Arion D et al (2006) Correlation of transcriptome profile with electrical activity in temporal lobe epilepsy. *Neurobiol Dis* 22, 374–387
69 Marciano PG et al (2002) Expression profiling following traumatic brain injury: a review. *Neurochem Res* 27, 1147–1155
70 Matzilevich DA et al (2002) High-density microarray analysis of hippocampal gene expression following experimental brain injury. *J Neurosci* Res 67, 646–663
71 Dash PK et al (2004) A molecular description of brain trauma pathophysiology using microarray technology: an overview. *Neurochem Res* 29, 1275–1286
72 Moore DF et al (2005) Using peripheral blood mononuclear cells to determine a gene expression profile of acute ischemic stroke: a pilot investigation. *Circulation* 111, 212–221
73 Tang Y et al (2006) Gene expression in blood changes rapidly in neutrophils and monocytes after ischemic stroke in humans: a microarray study. *J Cereb Blood Flow Metab* 26, 1089–1102
74 Kitano H (2002) Systems biology: a brief overview. *Science* 295, 1662–1664
75 Morel NM et al (2004) Primer on medical genomics. Part XIV: Introduction to systems biology – a new approach to understanding disease and treatment. *Mayo Clin Proc* 79, 651–658
76 Somogyi R et al (2006) Advanced data mining and predictive modeling at the core of personalized medicine. In: Paton & McNamara (eds): *Multidisciplinary Approaches to Theory in Medicine*, Vol. 3 Elsevier, Amsterdam, 165–192

Aspects of gene expression in B cell lymphomas

Enrico Tiacci[1], *Verena Brune*[1,2] *and Ralf Küppers*[1]

[1]Institute for Cell Biology (Tumor Research), Medical School, University of Duisburg-Essen, 45122 Essen, Germany;
[2]Institute of Pathology, University of Frankfurt/Main, 60590 Frankfurt, Germany

Abstract

Gene expression profiling has been used extensively for the analysis of human lymphomas. Besides the analysis of whole tissue or blood samples, genechip studies have also been applied to lymphoma cells isolated by cell sorting from cell suspensions or by laser microdissection from tissue sections. Such studies were successfully applied to clarify the cellular origin of lymphomas by comparison to normal B cell subsets, to gain insights into pathogenetic mechanisms, to develop molecular classifiers for differential diagnosis, to identify so far unrecognized subgroups among current lymphoma entities, and to establish predictors of prognosis. In our own work, we initially studied gene expression of Hodgkin lymphoma cell lines and revealed, for example, a global downregulation of the B cell gene expression programme and an aberrant expression of multiple receptor tyrosine kinases in these cells. We are now studying gene expression profiles from microdissected tumour cells of Hodgkin lymphoma.

Generation of gene expression profiles from normal and malignant B cells

High quality RNA is one of the most important requirements for the successful performance of genome-wide large-scale gene expression analysis. Peripheral blood lymphocytes or cells from cell suspensions of fresh biopsy specimens can be transferred alive directly into RNA lysis buffer, and the RNA quality as well as the original transcript levels will be preserved best. The quality of RNA isolated from tissue sections can differ considerably depending on the period of time from taking the biopsy to freezing or fixation, the type of fixative used and the time of storage for fixed and paraffin-embedded tissue. Therefore, it is highly recommended to test the RNA quality of tissues before using them for gene expression studies.

If no isolation of particular cells from the tissue is needed, RNA can be directly isolated from frozen tissue sections or from sections of paraffin-embedded material after extraction of paraffin. However, for the isolation of distinct cells or particular histological structures from tissue sections by microdissection, histological staining is required. Most standard histological staining procedures such as hematoxylin &

Microarrays in Inflammation, edited by Andreas Bosio and Bernhard Gerstmayer

eosin, methyl green, cresyl violet and nuclear Fast Red are suitable for the isolation of high-quality RNA if performed under RNase-free conditions. For RNase-rich tissues such as pancreas or spleen the use of a modified cresyl violet staining with short incubation times is recommended, since all solutions contain high concentrations of ethanol, which inhibits endogenous RNase activity (protocol available from P.A.L.M. Microlaser Technology, Bernried, Germany). Precise discrimination of cell types in conventionally stained sections is, however, limited. To identify cells not recognizable by morphology alone, immunohistochemical stainings may enable identification and isolation of these cells. Standard immunohistochemical staining protocols usually require several hours of incubation in aqueous solutions, resulting in significant degradation and loss of RNA mainly through activation of tissue RNases as well as other factors. Therefore, rapid immunohistochemical as well as immunofluorescence staining techniques were established for the use of laser microdissection and subsequent RT-PCR [1–4]. With single-step PCR it is possible to amplify fragments of more than 600 bp length from less than 500 immunostained and microdissected cells [2]. However, none of these immunohistochemical or immunofluorescence protocols have achieved wide acceptance for microdissection in combination with microarray analysis for global gene expression studies, likely because RNA degradation and loss after immunohistochemical or immunofluorescence staining is too severe for the complex requirements of RNA quality for microarray analysis.

Whole tissue analysis has been performed in many studies of global gene expression in lymphomas, causing the problem that it can be difficult to precisely identify the cellular origin of differentially expressed genes, as the lymphoma cells are always admixed with infiltrating non-tumour cells. Whole tissue analysis might, however, be suitable for prognostic purposes (see below), and to gain an initial insight into specific features of the lymphoma microenvironment, as exemplarily shown for follicular lymphomas [5]. To gain insights into the biology of lymphoma cells it is indispensable to isolate these cells. Lymphoma cells can be isolated by magnetic-activated cell separation (MACS), fluorescence-activated cell sorting (FACS) or laser microdissection. Microdissection of cells with a hydraulic apparatus and glass capillaries, which was often used previously for DNA studies [6], is not suitable for RNA analysis, as the long time the sections are covered by aqueous buffer during the microdissection leads to unacceptable loss of RNA. For laser microdissection, the sections are immediately dried after staining, reducing further RNA degradation or loss. Nevertheless, the RNA quality of the dry sections is reduced after a longer time, so that individual sections should be used not longer than six hours for laser microdissection (unpublished observation). If samples are snap-frozen for microdissection, the gene expression pattern of the cells should be fully retained. Moreover, microdissection has the advantage that the histological location of the cells of interest is known. Isolation of lymphoma cells by MACS and/or FACS from cell suspensions has the advantage that the live cells have optimal RNA quality, and that

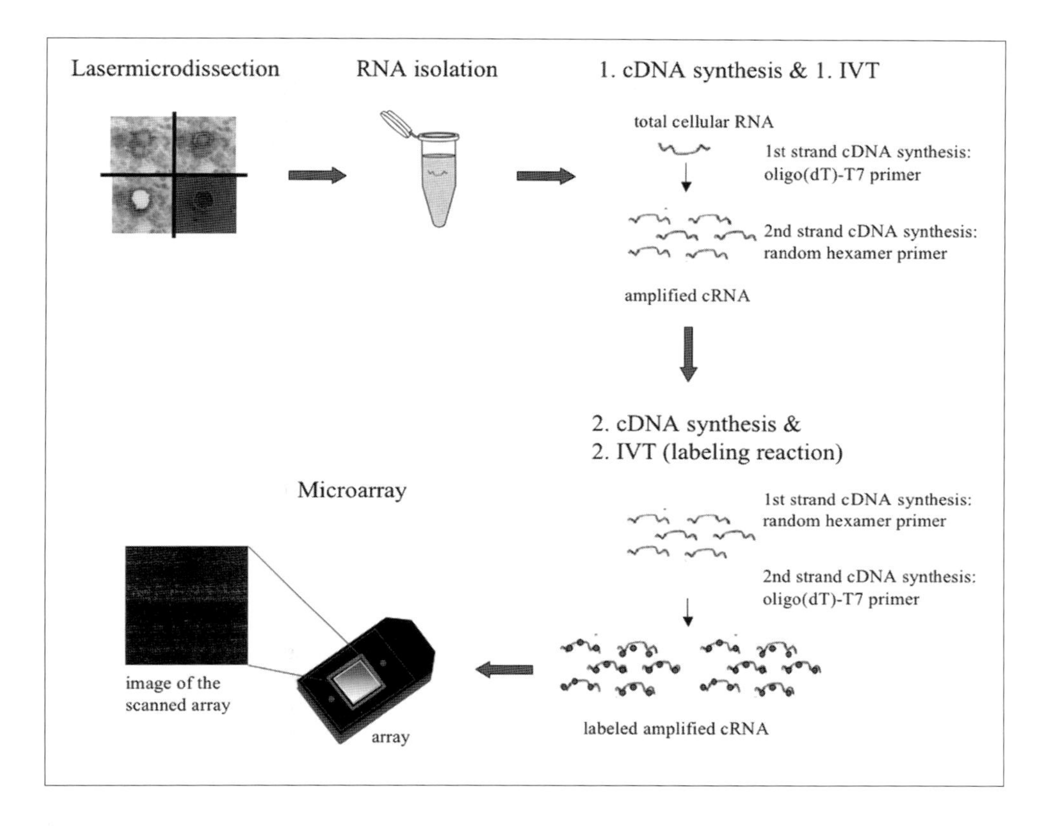

Figure 1
Schematic overview of the two-rounded T7 RNA polymerase-based in vitro *transcription (IVT) amplification protocol. After laser microdissection of the cells of interest the total cellular RNA is isolated and amplified with a two-rounded T7 RNA polymerase-based protocol. In a first 1st strand cDNA synthesis a T7 promoter sequence is incorporated using the oligo (dT)-T7 primer. The first 2nd strand cDNA synthesis utilizing random hexamer primers yields double-stranded cDNA. Following cDNA synthesis, IVT using T7 RNA polymerase yields antisense RNA (cRNA). In the second round of amplification, random hexamer primers are used for the second 1st stand cDNA synthesis followed by the second 2nd strand cDNA synthesis utilizing oligo (dT) primers. The second IVT is not only used to amplify RNA but also to label the cRNA by concomitant incorporation of biotin-modified ribonucleotides for subsequent hybridisation to microarrays.*

multiple markers can be used for the identification and isolation of the malignant cells. As the cells are still alive during cell sorting procedures, the cell sorting should be carried out as quickly as possible and the various experimental steps should be performed with cooling of the cells, to keep changes in gene expression as small as possible.

The standard T7 RNA polymerase-based amplification protocol with one round of *in vitro* transcription (IVT) requires 5 μg total RNA as starting material. Recent technical improvements allow for the use of only 1 μg of total RNA. However, 1 μg of total RNA still corresponds to approximately 10^5 to 10^6 cells. For the analysis of rare cell populations the use of a standard amplification protocol is hence not applicable. Therefore, the development of reproducible and reliable two-rounded T7 RNA polymerase-based IVT amplification protocols was an important achievement and has allowed the genome-wide gene expression analysis of small amounts of cells (Fig. 1) [7–13]. Even global gene expression analysis using formalin-fixed and paraffin-embedded (FFPE) material appears to be feasible using a two-round IVT amplification protocol and a special type of genechips [14]. However, amplification procedures risk distorting the initial transcript levels. Loss of signal strength for probes located relatively 5' in the mRNA is a characteristic feature of all linear amplification methods. The shortening of the cRNA fragments is mainly due to inefficient reverse transcription and the use of random hexamer primers in these reactions. This inevitably leads to loss of signals for targets located relatively 5' of the gene sequence. The use of Affymetrix GeneChip Human X3P arrays could, in part, overcome the problem of loss of information caused by the shortening of cRNA fragments, since this array was designed to focus on interrogated sequences located even closer to the 3' end of the transcripts compared to the standard Affymetrix design strategy, which selects probe sets mostly within the region of 600 base pairs proximal to the 3' ends. Due to the generation of systematic errors in the two-round protocol, it is mandatory that all samples included in the analysis be processed in the same way, thereby partly reducing biases generated by the experimental protocol.

As an alternative to *in vitro* transcription-based amplification protocols, also PCR- and strand displacement-based approaches have been developed for gene expression profiling of single or few cells [15–18]. A detailed description of one of these approaches is presented in the chapter by C. Klein of this volume.

Insights into B cell lymphoma biology and pathogenesis by gene expression profiling

Gene expression profiling studies have been performed for nearly all types of human B cell lymphomas in recent years. The aims of such studies were manifold: to develop diagnostic signatures, to establish predictors for prognosis, to identify potential novel subgroups among lymphomas, to gain insights into mechanisms involved in lymphoma pathogenesis, to identify the cellular origin of specific types of lymphomas, and to identify potential novel targets for therapy (Tab. 1). In the following, outstanding investigations addressing these issues are discussed.

In the first large study of gene expression profiling in human B cell lymphomas diffuse large B cell lymphomas (DLBCL) were analysed in comparison to normal

Table 1 - Applications of microarray studies in lymphomas and leukaemias

Aims	Examples	References
Development of prognostic classifier	mantle cell lymphoma, follicular lymphoma	[5, 27]
Development of diagnostic classifier	Burkitt lymphoma	[25, 26]
Identification of novel markers for diagnosis or prognosis	B-CLL (ZAP-70), HCL (annexin-1)	[31, 33]
Identification of novel lymphoma subgroups	diffuse large B cell lymphoma (ABC-DLBCL *versus* GC-DLBCL *versus* PMBCL)	[19, 20, 46, 47]
Identification of histogenetic origin of lymphoma cells	B-CLL, HCL, primary effusion lymphomas	[31–33, 48]
Identification of pathogenetic mechanisms	ABC-DLBCL (NFκB), HL (RTK)	[21, 40]
Identification of gene expression patterns determined by genetic aberrations	multiple myeloma, B cell acute lymphoblastic leukaemia	[28-30]
Identification of novel targets for therapy	ABC-DLBCL (NFκB), fatal/refractory DLBCL (PKCβ) HL (RTK)?	[21, 22, 40]

Abbreviations: ABC-DLBCL, activated B cell-like diffuse large B cell lymphoma; B-CLL, B cell chronic lymphocytic leukaemia; GC-DLBCL, germinal centre B cell-like diffuse large B cell lymphoma; HCL, hairy cell leukaemia; PMBCL, primary mediastinal B cell lymphoma; RTK, receptor tyrosine kinase.

lymphocyte populations and two types of low-grade B cell lymphomas [19]. The DLBCL expression profile was different from that of the other normal or malignant B cells in unsupervised hierarchical clustering. Most importantly, however, two subsets of DLBCL were identified, one expressing many genes typical for normal germinal centre (GC) B cells and one resembling *in vitro*-activated B cells. This led to the designation of GC B-like DLBCL (GC-DLBCL) and activated B-like DLBCL (ABC-DLBCL). These two subsets also showed different clinical behaviour after CHOP (i.e., cyclophosphamide, adriamycin, vincristine, prednisone) or CHOP-like chemotherapy, as patients with GC-DLBCL had a significantly better prognosis. This finding was extended and confirmed in a subsequent larger study with more than 200 patients, in which also a third, heterogeneous group of DLBCL was identified [20]. A comparison of genes distinguishing the GC- and ABC-DLBCL revealed that the latter type expressed many genes which are typical targets of the transcription factor NFκB [21]. It was indeed confirmed that ABC-DLBCL are characterized by constitutive activity of NFκB [21]. As cell lines derived from ABC-DLBCL showed propensity for apoptosis upon inhibition of NFκB, this might offer novel treatment options for this type of DLBCL. In another approach, an outcome predictor was developed for DLBCL treated with CHOP-based chemotherapy [22], which highlighted protein kinase C beta (PKCβ) as over-expressed in fatal/refractory DLBCL and as a promising candidate for targeted therapy in a subsequent clinical trial [23]. By using various clustering methods, three distinct subsets of DLBCL were identified, but the genes included in this model overlapped only partially with the ABC- *versus* GC-DLBCL defining genes. Based on prominent genes defining the three robust DLBCL subgroups, they were termed "oxidative phosphorylation", "B cell receptor/proliferation" and "host response" [24]. Thus, it appears that the tumour microenvironment and the host inflammatory response represent defining features for DLBCL subsets.

The distinction between DLBCL and Burkitt lymphoma may be difficult. This is however, clinically important as the two lymphomas require different treatments. Therefore, two groups performed gene expression studies to develop molecular classifiers for the differential diagnosis of DLBCL and Burkitt lymphoma [25, 26]. Both groups were able to define such classifiers that allowed a better distinction between these types of lymphomas. The Burkitt lymphoma classifier included high level of expression of c-myc, a gene that is affected by chromosomal translocations in Burkitt lymphoma, and low level of expression of NFκB target genes [25, 26].

Patients with mantle cell lymphoma have a bad prognosis with a median survival of about three years, but there is nevertheless a significant variation, with some patients dying after less than one year and others surviving for more than ten years. To better understand this variation at the level of gene expression, a genechip analysis of mantle cell lymphomas was performed [27]. A gene signature (i.e. a set of coordinately regulated genes) was identified that was associated with a worse prognosis. Interestingly, more than half of the genes defining this prediction signature belonged to proliferation-associated genes that are highly expressed in proliferating cells, indicating that cell cycle dysregulation plays a central role in mantle cell lymphoma pathogenesis.

In a similar study of nearly 200 cases of follicular lymphomas, it turned out that two gene signatures that best predict survival of the patients are mainly composed of genes not expressed by the lymphoma cells themselves but by other cells in the tumour microenvironment [5]. One of the signatures was mainly composed of genes expressed by T cells and macrophages, and the other signature included genes preferentially expressed by macrophages and/or dendritic cells. Importantly, the signatures were not merely a measure of the frequency of infiltrating cells, but more likely reflected a specific activation or differentiation stage of the cells. The strong impact of the cellular microenvironment on the prognosis of patients affected by follicular lymphoma presumably reflects the participation of the non-malignant immune cells mentioned above in the pathogenesis and biology of this lymphoma.

In multiple myeloma (MM) and pediatric B cell acute lymphoblastic leukaemia (B-ALL), gene expression profiling highlighted another important aspect of the pathogenesis of B cell derived neoplasms [28–30]. Unsupervised gene expression analyses of hundreds of MM and B-ALL cases revealed that both are characterized by significant transcriptional heterogeneity, which is largely dictated by those mutually exclusive genetic aberrations known to play a key role in the development of these neoplasms, i.e. structural chromosomal aberrations (such as reciprocal translocations deregulating the expression of MAF transcription factors and D-type cyclins in MM, or creating fusion genes like E2A-PBX1, TEL-AML1 and BCR-ABL in B-ALL), as well as numerical ones (such as hyperdiploidy in both MM and B-ALL). This suggests that even a single key genetic aberration can have a dramatic "avalanche" effect in the cell where it occurs, shaping its transcriptional profile to a considerable extent.

Gene expression profiling has also proven to be useful to shed light on the histogenesis of B cell malignancies, as exemplified by studies performed on B cell chronic lymphocytic leukaemia (B-CLL) and hairy cell leukaemia (HCL), two low grade neoplasms whose cellular origin has long been a matter of debate [31-33]. In fact, B-CLL comprises, in almost equal proportions, cases that show somatically mutated immunoglobulin (Ig) genes and cases that do not, with the formers having (for largely unknown reasons) a better prognosis than the latter. Conversely, in most of the other mature B-cell lymphomas, all or the overwhelming majority of cases are either mutated or unmutated [34]. As the GC of secondary lymphoid follicles represents the main environment where Ig mutations occur in normal B cells to increase the affinity for the stimulating antigen [6] (and also where the Ig isotype is changed to modify its effector function), it was relatively straightforward to assign a pre-GC origin to the unmutated lymphomas and a GC or post-GC origin to the mutated ones [35]. Expression analysis of single markers does not help to clarify the B-CLL cell of origin, because B-CLL is positive for markers of both naïve B cells (such as CD5 and CD23) and post-GC memory B-cells (e.g. CD27). Regarding HCL, the problem in identifying its normal cellular counterpart was not the heterogeneous mutational status of its Ig genes, as most HCL cases are, in fact, mutated, but rather – and again in contrast with the other B cell neoplasms – the frequent coexpression of multiple and clonally related Ig isotypes (IgM, IgD, IgG and/or IgA) and certain features of GC B cells, e.g. some expression of activation-induced cytidine deaminase, and some level of intraclonal Ig genes diversification. By comparing, on a genome-wide rather than on a single-gene basis, the expression profiles of B-CLL and HCL with that of the various normal B cell subsets (naïve B cells, GC B cells, memory B cells and plasma cell lines), it was possible to find that B-CLL irrespective of their Ig mutational status and HCL are both significantly more similar to post-GC memory B cells than to the other B cell populations [31–33].

Establishing a putative cell of origin by means of gene expression profiling also allowed highlighting the differences between B-CLL and HCL and their potential normal counterpart, which are likely introduced by the respective (and, thus far, largely unknown) underlying transforming events. For example, leukemic B-CLL cells surprisingly show a less proliferative and more anti-apoptotic transcriptional phenotype as compared to normal memory B cells, which are already regarded as a relatively quiescent population. HCL cells, on the other hand, display a different profile of chemokine receptors, adhesion molecules and inhibitors of matrix metalloproteinases that might explain their propensity to disseminate in blood-related compartments without significantly entering the lymphoid tissue parenchyma, as normal memory B cells do instead.

The same studies on B-CLL and HCL [31, 33] are also paradigmatic of the potential that expression profiling has in translating its genome-wide, complex results into simple and widely applicable, single-gene based assays as valuable prognostic or diagnostic tools. Indeed, among the few genes differentially expressed between mutated and unmutated B-CLL cases, there is ZAP70, an intracellular signaling protein whose expression in B-CLL is largely concordant with the absence of Ig mutations [33]. Therefore, ZAP70 has been successfully used for prognostic stratification of B-CLL patients using flow cytometry or immunohistochemistry [36, 37]. Annexin-1, one of the 81 HCL-specific genes highlighted by gene expression profiling [31], turned out to be diagnostically useful in routine immunohistochemistry for differentiating HCL from other chronic B cell lymphoproliferative disorders, such as splenic lymphoma with villous lymphocytes and HCL-variant, which have a similar clinical presentation and immunophenotype but do not respond well to the drugs used in HCL patients [38].

Differential gene expression in Hodgkin lymphoma

In classical Hodgkin lymphoma (HL) the tumour cells, which are called Hodgkin and Reed-Sternberg (HRS) cells, usually account for less than 1% of cells in the lymphoma. This has severely hampered their molecular analysis, and much work has consequently been performed with HL-derived cell lines. We performed a global analysis of differential gene expression in HRS cells by analysing four HL cell lines in comparison to the major subsets of normal human mature B cells (naïve, GC and memory) and to cell lines or primary biopsy material from human B cell Non-Hodgkin lymphomas [39]. In this analysis, we identified an HL-specific gene expression pattern that included 27 genes showing a unique, increased expression in the HRS cells. Twenty-three of these genes were not previously known to be expressed by HRS cells, including transcription factors (ABF1, GATA3 and Nrf3) and the tyrosine kinase Fer. These genes may play important roles in the pathogenesis of HL.

As receptor tyrosine kinases (RTKs) play a central role in the pathogenesis of numerous malignancies, we also screened the gene expression data for expression of these genes. Indeed, we observed transcription of several RTKs, each by at least two of the four HL cell lines [40]. For five RTKs (PDGFRA, DDR2, EPHB1, RON, TRKA), expression by HRS cells was seen in 30-75% of HL cases by immunohistochemistry. Such a coexpression of multiple RTKs has not been observed before in any malignancy. Expression of these genes correlated with elevated overall phosphotyrosine levels in HRS cells, and for two of the RTKs, for which phospho-specific antibodies were available, phosphorylation of the RTKs indicated their activation [40]. This activity is presumably not due to activating mutations, but is ligand-induced, as we detected expression of ligands for four of the RTKs in either the HRS cells or in other cells in the HL microenvironment.

Among the genes downregulated in HL cell lines in comparison to the other B cell lymphomas and normal B cells, we recognized a large number of genes that play key roles in the function of B cells. Indeed, a thorough analysis of the downregulated genes revealed a global loss of the B cell phenotype [41]. This is not a particular feature of the cell lines, as we confirmed loss of expression for nine of the B cell markers by immunohistochemistry also for HRS cells in HL biopsy specimens. Notably, the downregulation of B cell-typical genes did not include genes involved in antigen presentation and interaction with T helper cells (e.g. CD80, CD86 and MHC class II), indicating that antigen-presenting functions are still important for HRS cells. As HRS cells are presumably derived from GC B cells that lost the capacity to express a functional B cell receptor [42], we speculated that the lost B cell phenotype may be linked to the cellular origin of the HRS cells, and represent a means to evade the apoptotic program that normally leads to apoptosis of GC B cells lacking expression of a B cell receptor. The dowregulation of many B cell-specific genes in HRS cells is likely due to several mechanisms, including aberrant expression of suppressors of B cell-specific genes, i.e. ID2 and ABF1 [43].

Although valuable results have been obtained by gene expression profiling of classical HL cell lines, it is likely that cultured HRS cells do not reflect primary HRS cells in all biological aspects. Indeed, HL lines were derived from sites (such as bone marrow, pleural effusion and peripheral blood) [44] which are very rarely involved by HL and where HRS cells lost one of their potential key features, i.e. the interaction with the prominent inflammatory cellular background surrounding HRS cells in the lymph node. Therefore, we have also generated genome-wide expression profiles (Affymetrix chip U133 Plus 2.0) from primary HRS cells, which were singly microdissected from lymph node biopsies (1000-2000 cells per case) of classical HL cases. The preliminary results (unpublished) of this study indeed show a highly differential regulation of more than 1000 genes between primary and cultured HRS cells. While many of the genes upregulated by the latter obviously belong to a proliferation signature that tumour cells had to acquire to adapt themselves to growth in suspension (first in the patient, and then *in vitro*), a good number of genes are also upregulated by primary HRS cells, including a variety of chemokines, chemokine receptors and extracellular matrix remodeling factors, thus pointing to an influence on/from the lymph node microenvironment which is largely lost in HL cell lines. Profiling primary HRS cells therefore has the potential to unravel new aspects of HL pathogenesis that analysis of HL cell lines might not be able to elucidate.

Concluding remarks

Numerous mRNA expression profiling studies of human B cell lymphomas have demonstrated the power of this technique to gain novel insights into the biology and pathogenesis of these malignancies, to identify novel disease subtypes and to develop prognostic classifiers. With the recognition that the expression of many protein-encoding genes is often regulated at the post-transcriptional level by small, non-coding RNAs, so called microRNAs, it has become clear that, for a comprehensive understanding of the gene expression pattern of a lymphoma, one has to take the transcriptome of mRNAs and miRNAs into consideration [45]. An additional future challenge and prospect lies in the combination of global gene expression studies with other genome-wide analyses of human lymphomas, in particular those characterizing DNA methylation status and genomic imbalances. With the help of such studies, the functional relationship between gene expression and genomic structures and aberrations can be determined at an unprecedented level of resolution.

Acknowledgements
Own work discussed in this review was supported in part by grants from the Deutsche Forschungsgemeinschaft (Ku1315/5-2), the Deutsche Krebshilfe, Mildred Scheel-Stiftung, and the German José Carreras Foundation (fellowship F05/01 to ET).

References

1 Fend F, Emmert-Buck MR, Chuaqui R, Cole K, Lee J, Liotta LA, Raffeld M (1999) Immuno-LCM: laser capture microdissection of immunostained frozen sections for mRNA analysis. *Am J Pathol* 154: 61–66
2 Fend F, Raffeld M (2000) Laser capture microdissection in pathology. *J Clin Pathol* 53: 666–672
3 Fink L, Kinfe T, Seeger W, Ermert L, Kummer W, Bohle RM (2000) Immunostaining for cell picking and real-time mRNA quantitation. *Am J Pathol* 157: 1459–1466
4 Fink L, Kinfe T, Stein MM, Ermert L, Hanze J, Kummer W, Seeger W, Bohle RM (2000)

Immunostaining and laser-assisted cell picking for mRNA analysis. *Lab Invest* 80: 327–333

5 Dave SS, Wright G, Tan B, Rosenwald A, Gascoyne RD, Chan WC, Fisher RI, Braziel RM, Rimsza LM, Grogan TM et al (2004) Prediction of survival in follicular lymphoma based on molecular features of tumor-infiltrating immune cells. *N Engl J Med* 351: 2159–2169

6 Küppers R, Zhao M, Hansmann ML, Rajewsky K (1993) Tracing B cell development in human germinal centres by molecular analysis of single cells picked from histological sections. *EMBO J* 12: 4955–4967

7 Kenzelmann M, Klaren R, Hergenhahn M, Bonrouhi M, Grone HJ, Schmid W, Schutz G (2004) High-accuracy amplification of nanogram total RNA amounts for gene profiling. *Genomics* 83: 550–558

8 Luo L, Salunga RC, Guo H, Bittner A, Joy KC, Galindo JE, Xiao H, Rogers KE, Wan JS, Jackson MR et al (1999) Gene expression profiles of laser-captured adjacent neuronal subtypes. *Nat Med* 5: 117–122

9 Luzzi V, Holtschlag V, Watson MA (2001) Expression profiling of ductal carcinoma *in situ* by laser capture microdissection and high-density oligonucleotide arrays. *Am J Pathol* 158: 2005–2010

10 Luzzi V, Mahadevappa M, Raja R, Warrington JA, Watson MA (2003) Accurate and reproducible gene expression profiles from laser capture microdissection, transcript amplification, and high density oligonucleotide microarray analysis. *J Mol Diagn* 5: 9–14

11 McClintick JN, Jerome RE, Nicholson CR, Crabb DW, Edenberg HJ (2003) Reproducibility of oligonucleotide arrays using small samples. *BMC Genomics* 4: 4

12 Ohyama H, Zhang X, Kohno Y, Alevizos I, Posner M, Wong DT, Todd R (2000) Laser capture microdissection-generated target sample for high-density oligonucleotide array hybridization. *Biotechniques* 29: 530–536

13 Wilson CL, Pepper SD, Hey Y, Miller CJ (2004) Amplification protocols introduce systematic but reproducible errors into gene expression studies. *Biotechniques* 36: 498–506

14 Frank M, Döring C, Metzler D, Eckerle S, Hansmann ML (2007) Global gene expression profiling of formalin-fixed paraffin-embedded tumor samples: a comparison to snap-frozen material using oligonucleotide microarrays. *Virchows Arch* 450: 699–711

15 Hartmann CH, Klein CA (2006) Gene expression profiling of single cells on large-scale oligonucleotide arrays. *Nucleic Acids Res* 34: e143

16 Kurimoto K, Yabuta Y, Ohinata Y, Ono Y, Uno KD, Yamada RG, Ueda HR, Saitou M (2006) An improved single-cell cDNA amplification method for efficient high-density oligonucleotide microarray analysis. *Nucleic Acids Res* 34: e42

17 Dafforn A, Chen P, Deng G, Herrler M, Iglehart D, Koritala S, Lato S, Pillarisetty S, Purohit R, Wang M et al (2004) Linear mRNA amplification from as little as 5 ng total RNA for global gene expression analysis. *Biotechniques* 37: 854–857

18 Singh R, Maganti RJ, Jabba SV, Wang M, Deng G, Heath JD, Kurn N, Wangemann

P (2005) Microarray-based comparison of three amplification methods for nanogram amounts of total RNA. *Am J Physiol Cell Physiol* 288: C1179–1189

19 Alizadeh AA, Eisen MB, Davis RE, Ma C, Lossos IS, Rosenwald A, Boldrick JC, Sabet H, Tran T, Yu x et al (2000) Distinct types of diffuse large B-cell lymphoma identified by gene expression profiling. *Nature* 403: 503–511

20 Rosenwald A, Wright G, Chan WC, Connors JM, Campo E, Fisher RI, Gascoyne RD, Muller-Hermelink HK, Smeland EB, Giltnane JM et al (2002) The use of molecular profiling to predict survival after chemotherapy for diffuse large-B-cell lymphoma. *N Engl J Med* 346: 1937–1947

21 Davis RE, Brown KD, Siebenlist U, Staudt LM (2001) Constitutive nuclear factor kappaB activity is required for survival of activated B cell-like diffuse large B cell lymphoma cells. *J Exp Med* 194: 1861–1874

22 Shipp MA, Ross KN, Tamayo P, Weng AP, Kutok JL, Aguiar RC, Gaasenbeek M, Angelo M, Reich M, Pinkus GS et al (2002) Diffuse large B-cell lymphoma outcome prediction by gene-expression profiling and supervised machine learning. *Nat Med* 8: 68–74

23 Robertson MJ, Kahl BS, Vose JM, de Vos S, Laughlin M, Flynn PJ, Rowland K, Cruz JC, Goldberg SL, Musib L et al (2007) Phase II study of enzastaurin, a protein kinase C beta inhibitor, in patients with relapsed or refractory diffuse large B-cell lymphoma. *J Clin Oncol* 25: 1741–1746

24 Monti S, Savage KJ, Kutok JL, Feuerhake F, Kurtin P, Mihm M, Wu B, Pasqualucci L, Neuberg D, Aguiar RC et al (2005) Molecular profiling of diffuse large B-cell lymphoma identifies robust subtypes including one characterized by host inflammatory response. *Blood* 105: 1851–1861

25 Dave SS, Fu K, Wright GW, Lam LT, Kluin P, Boerma EJ, Greiner TC, Weisenburger DD, Rosenwald A, Ott G et al (2006) Molecular diagnosis of Burkitt's lymphoma. *N Engl J Med* 354: 2431–2442

26 Hummel M, Bentink S, Berger H, Klapper W, Wessendorf S, Barth TF, Bernd HW, Cogliatti SB, Dierlamm J, Feller AC et al (2006) A biologic definition of Burkitt's lymphoma from transcriptional and genomic profiling. *N Engl J Med* 354: 2419–2430

27 Rosenwald A, Wright G, Wiestner A, Chan WC, Connors JM, Campo E, Gascoyne RD, Grogan ™, Müller-Hermelink HK, Smeland EB et al (2003) The proliferation gene expression signature is a quantitative integrator of oncogenic events that predicts survival in mantle cell lymphoma. *Cancer Cell* 3: 185–197

28 Ross ME, Zhou X, Song G, Shurtleff SA, Girtman K, Williams WK, Liu HC, Mahfouz R, Raimondi SC, Lenny N et al (2003) Classification of pediatric acute lymphoblastic leukemia by gene expression profiling. *Blood* 102: 2951–2959

29 Yeoh EJ, Ross ME, Shurtleff SA, Williams WK, Patel D, Mahfouz R, Behm FG, Raimondi SC, Relling MV, Patel A et al (2002) Classification, subtype discovery, and prediction of outcome in pediatric acute lymphoblastic leukemia by gene expression profiling. *Cancer Cell* 1: 133–143

30 Zhan F, Huang Y, Colla S, Stewart JP, Hanamura I, Gupta S, Epstein J, Yaccoby S,

Sawyer J, Burington B et al (2006) The molecular classification of multiple myeloma. *Blood* 108: 2020–2028

31 Basso K, Liso A, Tiacci E, Benedetti R, Pulsoni A, Foa R, Di Raimondo F, Ambrosetti A, Califano A, Klein U et al (2004) Gene expression profiling of hairy cell leukemia reveals a phenotype related to memory B cells with altered expression of chemokine and adhesion receptors. *J Exp Med* 199: 59–68

32 Klein U, Tu Y, Stolovitzky GA, Mattioli M, Cattoretti G, Husson H, Freedman A, Inghirami G, Cro L, Baldini L et al (2001) Gene expression profiling of B cell chronic lymphocytic leukemia reveals a homogeneous phenotype related to memory B cells. *J Exp Med* 194: 1625–1638

33 Rosenwald A, Alizadeh AA, Widhopf G, Simon R, Davis RE, Yu X, Yang L, Pickeral OK, Rassenti LZ, Powell J et al (2001) Relation of gene expression phenotype to immunoglobulin mutation genotype in B cell chronic lymphocytic leukemia. *J Exp Med* 194: 1639–1647

34 Küppers R (2005) Mechanisms of B-cell lymphoma pathogenesis. *Nat Rev Cancer* 5: 251–262

35 Küppers R, Klein U, Hansmann M-L, Rajewsky K (1999) Cellular origin of human B-cell lymphomas. *N Engl J Med* 341: 1520–1529

36 Crespo M, Bosch F, Villamor N, Bellosillo B, Colomer D, Rozman M, Marce S, Lopez-Guillermo A, Campo E, Montserrat E (2003) ZAP-70 expression as a surrogate for immunoglobulin-variable-region mutations in chronic lymphocytic leukemia. *N Engl J Med* 348: 1764–1775

37 Rassenti LZ, Huynh L, Toy TL, Chen L, Keating MJ, Gribben JG, Neuberg DS, Flinn IW, Rai KR, Byrd JC et al (2004) ZAP-70 compared with immunoglobulin heavy-chain gene mutation status as a predictor of disease progression in chronic lymphocytic leukemia. *N Engl J Med* 351: 893–901

38 Falini B, Tiacci E, Liso A, Basso K, Sabattini E, Pacini R, Foa R, Pulsoni A, Dalla Favera R, Pileri S (2004) Simple diagnostic assay for hairy cell leukaemia by immunocytochemical detection of annexin A1 (ANXA1). *Lancet* 363: 1869–1870

39 Küppers R, Klein U, Schwering I, Distler V, Bräuninger A, Cattoretti G, Tu Y, Stolovitzky GA, Califano A, Hansmann ML et al (2003) Identification of Hodgkin and Reed-Sternberg cell-specific genes by gene expression profiling. *J Clin Invest* 111: 529–537

40 Renné C, Willenbrock K, Küppers R, Hansmann ML, Bräuninger A (2005) Autocrine and paracrine activated receptor tyrosine kinases in classical Hodgkin lymphoma. *Blood* 105: 4051–4059

41 Schwering I, Bräuninger A, Klein U, Jungnickel B, Tinguely M, Diehl V, Hansmann ML, Dalla-Favera R, Rajewsky K, Küppers R (2003) Loss of the B-lineage-specific gene expression program in Hodgkin and Reed-Sternberg cells of Hodgkin lymphoma. *Blood* 101: 1505–1512

42 Küppers R (2002) Molecular biology of Hodgkin's lymphoma. *Adv Cancer Res* 84: 277–312

43 Küppers R, Bräuninger A (2006) Reprogramming of the tumour B-cell phenotype in Hodgkin lymphoma. *Trends Immunol* 27: 203–205
44 Küppers R, Re D (2007) Nature of Reed-Sternberg and L&H cells, and their molecular biology in Hodgkin lymphoma. In: RT Hoppe, JO Armitage, V Diehl, PM Mauch, LM Weiss, (eds): *Hodgkin lymphoma*. Lippincott Williams & Wilkins, Philadelphia, 73–86
45 Lawrie CH (2007) MicroRNAs and haematology: small molecules, big function. *Br J Haematol* 137: 503–512
46 Rosenwald A, Wright G, Leroy K, Yu X, Gaulard P, Gascoyne RD, Chan WC, Zhao T, Haioun C, Greiner TC et al (2003) Molecular diagnosis of primary mediastinal B cell lymphoma identifies a clinically favorable subgroup of diffuse large B cell lymphoma related to Hodgkin lymphoma. *J Exp Med* 198: 851–862
47 Savage KJ, Monti S, Kutok JL, Cattoretti G, Neuberg D, De Leval L, Kurtin P, Dal Cin P, Ladd C, Feuerhake F et al (2003) The molecular signature of mediastinal large B-cell lymphoma differs from that of other diffuse large B-cell lymphomas and shares features with classical Hodgkin lymphoma. *Blood* 102: 3871–3879
48 Klein U, Gloghini A, Gaidano G, Chadburn A, Cesarman E, Dalla-Favera R, Carbone A (2003) Gene expression profile analysis of AIDS-related primary effusion lymphoma (PEL) suggests a plasmablastic derivation and identifies PEL-specific transcripts. *Blood* 101: 4115–4121

The genes behind rheumatology

Thomas Häupl[1], Andreas Grützkau[2], Bruno Stuhlmüller[1], Karl Skriner[1], Gerd Burmester[1] and Andreas Radbruch[2]

[1]Department of Rheumatology and Clinical Immunology, Charité-University Medicine, Berlin, Germany; [2]German Arthritis Research Center (DRFZ), Berlin, Germany

Abstract

Chronic inflammation is the central pathophysiology in rheumatic diseases. Blood contains most types of immune cells involved in this inflammation. Investigation of purified cell types has various advantages but also technical limitations compared to whole blood. Infiltration of immune cells into involved organs is a leading phenomenon and molecular processes in the inflamed tissue seem to change and increase by number and intensity compared to those in the blood. Many genes found differentially expressed in tissues are elevated as part of the infiltration, while only a smaller fraction becomes regulated upon inflammation. Gene products as biomarkers, which are affected by disease-specific pathomechanisms or by treatment, are of diagnostic importance. With the growing number of different biologics, markers indicative of different pathomechanisms are important for predicting drug responsiveness and treatment stratification. Finally, a comparison of pathological signatures between different types of diseases will contribute to identifying both similarities and differences in molecular processes and thus may provide insight into the driving mechanisms and etiology of rheumatic diseases.

Introduction

The aetiology of chronic inflammatory processes is unclear for most cases of rheumatic diseases. A wide range of immune processes, including adaptive immune responses directed against self-antigens, suggests that autoimmune pathomechanisms are important. In rheumatoid arthritis (RA), the most frequent entity of these chronic inflammatory diseases, the leading autoantigen is the immunoglobulin targeted by rheumatoid factor. A similar or even greater association with RA was found for antibodies to citrullinated peptides, which are protein modifications that occur, for example, during apoptosis, a frequent process in chronic inflammation. In particular, reactivity to citrullinated vimentin seems to be highly specific [1] and of prognostic relevance [2]. Still, why these self-antigens are targeted predominantly in RA remains unclear. In connective tissue diseases such as systemic lupus erythematosus or systemic sclerosis, other autoantigens are found involving

Microarrays in Inflammation, edited by Andreas Bosio and Bernhard Gerstmayer

predominantly nuclear antigens [3]. Autoantigens are products of one set of genes behind rheumatology that provide the targets of a misdirected immune response. While these autoantigenes are important markers for diagnostic application, there are currently no relevant strategies for therapeutic application arising from knowledge of these, except for depletion of antibodies by means of immune adsorption [4] or by unspecific B cell depletion [5, 6]. As demonstrated by the development of new biologic drugs, genes with increased expression in rheumatic diseases and involvement in the regulation of the molecular network of inflammation are of much more interest [7–9]. Innate as well as adaptive immune processes have been investigated and both arms of the immune system seem to be involved in these chronic inflammatory diseases. Interestingly, up to now the overlap between genes that are behind the autoantigens and genes that are regulated in the inflammatory response has not been investigated in detail. A literature search and own observations comparing differential gene expression in various cell types and tissues with an autoantigen screen in rheumatoid arthritis did not suggest that there is a relevant pathophysiologic link between over-expression of a gene and its recognition by the adaptive immune system as an autoantigen. While autoreactivity patterns have their value for diagnostic classification of the disease, most of them are not valuable markers for disease activity. Therefore, it is important to investigate the influence of the different drugs currently on the market not only to demonstrate their inhibitory potential and specific role in the pathophysiology of these diseases [10] but also to identify new diagnostic markers to monitor the course of the disease within a short timeframe. In this chapter, we will present an overview of the current status of identified genes in the network of inflammation and its change upon therapy.

Samples, compartments, and their impact on the gene selection process

For many practical reasons, blood is the most favourable type of sample used for clinical investigations. Therefore, many analyses have focused on gene expression in standard cell extracts of blood, especially peripheral blood mononuclear cells (PBMC). However, there are several limitations to interpreting such gene expression profiles. These include the risk of *in vitro* activation [11] as well as changes in gene expression related to differential cellular composition. Only purification and analysis of individual cell populations can help to allocate changes of gene expression to a particular cell type [12, 13]. In diseases such as rheumatoid arthritis with inflammation and destruction predominantly in the joint, differences in gene activation between blood and tissue compartments must be expected. Furthermore, in blood and tissue samples, there are changes in cellular composition which confound the interpretation of differential expression. Table 1 illustrates, on the basis of an at least 2-fold changed expression, the striking differences in gene expression

Table 1 - Differential expression of genes between different types of blood cells (A) and between rheumatoid arthritis (RA) and normal donors (ND) for 3 different cell types, whole blood and synovial tissue (B). Numbers of Affymetrix 133A probe sets were determined that revealed changed expression by more than 2-fold.

	A						B	
	Mo vs...	**PMN vs...**	**CD4 vs...**	**CD19 vs...**	**NK vs...**	**SF vs...**	**RAvsND**	**NDvsRA**
Mo	0	2599	1958	2578	3805	2554	71	17
PMN	2625	0	3112	3706	4818	3851	99	58
CD4	1853	2898	0	1899	3193	2762	52	61
CD19	1985	3020	1626	0	2907	2908		
NK	2060	2939	1786	1868	0	2868		
SFbl	2572	3426	2842	3519	4571	0		
blood							205	91
tissue							1129	1124

Mo: monocytes; PMN: granulocytes; CD4: $CD4^+$ T-cells; CD19: $CD19^+$ B-cells; NK: natural killer cells; SFbl: synovial fibroblasts; blood: whole blood purified from PAXgene samples; tissue: synovial biopsies

between different cell types purified from normal controls. In contrast, only small differences can be identified between a disease condition like rheumatoid arthritis and normal controls when investigating purified cell types from peripheral blood or whole blood. However, differences identified in inflamed as compared to normal tissue are at least 20-fold larger than those in blood. This may be in part explained by the infiltration of immune cells in the inflamed tissue, as this is associated with a substantial change in cellular composition and contributes the cell type specific differences shown in Table 1. A quantitative estimate of this influence is summarized in Figure 1. According to the large number of overlapping genes, only a quarter or less of the genes differentially expressed in the inflamed tissue may be truly regulated and may be informative for the molecular pathomechanisms, while other genes merely reflect a differential cell count. With this limitation in mind, subsequently summarized investigations on blood as well as on tissue samples must be interpreted carefully. Only little data are available on purified cell types, while the problem with samples consisting of cell mixtures has not been sufficiently addressed in most investigations published up to now. Finally, for many genes, differential expression only indicates a change in the capability to react upon appropriate stimulation but does not necessarily indicate that the pathways behind these genes are truly activated.

samples	probe sets with present calls	increased compared to normal tissue biopsies	overlap with increase in RA tissue biopsies	all overlaps
monocyte	6754	2568	674	904
granulocyte	6400	3201	507	
B-cell	7742	2097	490	
T-cell	6442	3132	473	
normal synovial tissue biopsy	6074	0	0	
RA synovial tissue biopsy	7126	1236		

Figure 1
A large fraction of genes differentially expressed in RA versus normal control tissue seem to reflect the differential cell count after infiltration of immune cells into synovial tissue. The first column gives the number of probe sets that were found to be expressed in the different cell types and tissues. The second column presents the number of probe sets increased in at least 50% of the samples of an individual cell type compared to normal synovial biopsies. These are the candidate genes related to infiltration. The third column demonstrates the overlap of these infiltration-related candidates with the 1236 probe sets increased in RA versus *normal control tissue. These add up to 904 probe sets or about 75% of all probe sets differentially expressed in RA* versus *normal synovium.*

Genes identified by investigations of blood cells

The most interesting results were found by several groups investigating PBMC or whole blood cells from patients with systemic lupus erythematosus [14-16]. While Rus et al. [17] focused on cytokine- and chemokine-related genes in a customized set of only 375 genes, and was able to distinguish SLE from controls with a predominance of genes increased in the family of the TNF/death receptor, the IL-1 cytokine family and IL-8 and its receptors, the other groups applied much larger arrays with up to 10,000 genes and identified an interferon signature. In addition to this interferon signature, Bennett et al. found an increase in a granulopoiesis profile in PBMC and Baechler et al. a decrease of typical lymphocyte related genes such as LCK and T cell receptor genes.

PBMC were also investigated in several arthritides including rheumatoid arthritis (RA), spondyloarthropathy (SpA), and psoriatic arthritis (PsA). Gu et al. published their comparative study on all three arthritides using a 588-gene microarray. Beside various activation and differentiation markers for monocyte/macrophages, they pointed at CXCR4/SDF-1 as a potential pro-inflammatory axis for RA, PsA and SpA [18]. Similarly, Bovin et al. suggested an altered phagocytic function when comparing PBMC of rheumatoid factor positive (RF+) and RF– patients with healthy controls [19]. Furthermore, they found no differences between both RA groups. Batliwalla et al. (20) also confirmed a phagocyte/monocyte related dominance that correlated with monocyte counts and suggested, *via* the differential expression of Grb2-associated binding protein (GAB2) and its involvement in the regulation of the phosphatase function, a link to the recently proposed genetic risk associated with a mutation in PTPN22 [21]. The attempt to differentiate between early and long standing RA by Olsen et al. [22] demonstrated not only that expression profiling enables detection of discriminative gene clusters but also that subgroups in early RA exist which overlap in part with response patterns induced by influenza vaccination or found in a subset of SLE patients. The genes were functionally grouped into the categories of immune/growth factors, metabolism, neuromuscular, and transcription, and also included the granulocyte associated gene CSF3 receptor. Most recent investigations of whole blood preparations generated with PAXgene technology identified a subgroup of RA patients with upregulation of interferon inducible genes [23]. The same samples were also compared to pathogen response programs [24]. The gene expression pattern of a subgroup of these patients was reminiscent of the response of macaques to poxvirus infection and revealed, for the promoter regions of over-expressed genes, an enrichment of binding sites for interferon activated transcription factors. In fact, expression of IFN-response genes was also increased and this subgroup, with activation of a host-pathogen response pattern, was further characterized by increased autoreactivity to citrullinated proteins.

In summary, these studies demonstrate that differences between rheumatic diseases and controls can be identified in gene expression activity of peripheral blood cells. However, to interpret the complex network of data generated by these analyses, reference signatures of stimulated cells, such as the IFN signature or pathogen response patterns, are essential to annotate the individual genes to a functional process. Furthermore, some typical cell-type related genes were found differentially expressed, which could indicate: i) impurity of the PBMC fraction, especially in inflamed conditions (granulocytes), ii) a shift in the cellular composition of the peripheral blood cells (lymphocytes), or iii) a true increase/decrease in gene activity related to the pathomolecular changes in some of the cell types of PBMC or whole blood.

To overcome the problem of mixed cell populations, which confound interpretation of expression profiles and predominantly represent expression in the major cell population or high copy number genes of minor populations, analysis of puri-

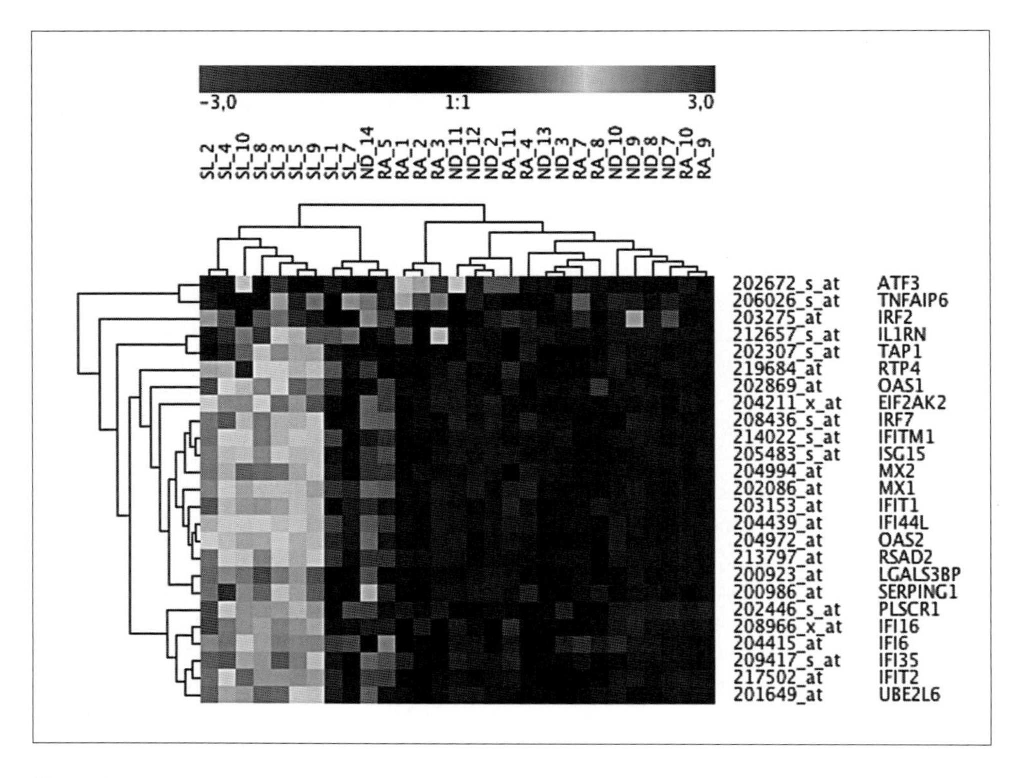

Figure 2
Expression of IFN-response genes in monocytes purified from ND (n=10), RA- (n=10) and SLE- (n=9) patients.

fied cell types is important [25]. Own efforts have focused on isolation techniques that avoid artificial activation of the cells [13, 26] as this is an important issue for subsequent expression profiling [27]. Phagocytic expression patterns have repeatedly been identified by several groups and in different rheumatic diseases when analyzing whole blood or PBMC. Earlier studies based on differential hybridization experiments showed a broad panel of genes upregulated in monocytes from patients with active RA [28]. Subsequent comparative analysis between RA, SpA, SLE and healthy controls by microarrays revealed a distinct activation pattern characteristic of each type of disease. In particular, SLE patients dominated with a typical IFN-type I response pattern which was completely different from the expression pattern in RA patients [13, 29]. While upregulation of the IFN-response genes as defined earlier [23] was remarkable in monocytes of SLE patients, there was no relevant overlap with RA gene regulation in monocytes of our patients. This indicates that the group of RA patients may be different or that substantial differences may exist for monocyte expression in RA and SLE concerning IFN response (Fig. 1). The latter would also suggest that data from different types of samples may vary and mixed cell populations may not necessarily represent a cellular response pattern common to all cell types.

Genes identified by investigation of tissues and tissue derived cells

Table 1 demonstrates that, in inflamed tissues, 5- to 10-fold higher numbers of differentially expressed genes may be identified. Figure 1 gives an estimate that about 3/4 of the differentially expressed genes may be related to the process of cell infiltration and do not necessarily reflect gene activation as part of the pathological process. This effect may play a substantial role in the observations of Lindberg et al. [30]. They addressed the issue of sampling heterogeneity by testing the effect of the sampling technique (open surgery *versus* arthroscopy) and by comparing different samples taken at the same time from different sites of the same knee joint. They demonstrated that major differences were found comparing samples from open surgery not only by microarray analysis but also by histological comparison, with variation in the intensity of inflammation. However, when comparing samples from arthroscopic procedures, changes were minor and analysis provided expression patterns that were unique for each patient. Own unpublished observations suggest that this variability is especially influenced by the size of the biopsy. While larger pieces taken during open surgery include deeper layers with less inflammation, smaller biopsies from macroscopically inflamed villae concentrate more on the high inflammation in the superficial layers of the synovium.

Nevertheless, important results have been achieved by microarray analysis of rheumatoid synovium. Van der Pouw Kraan et al. [31] investigated synovium of 15 RA patients on a 24000 cDNA microarray and was able to distinguish between two major subgroups of samples. One group revealed a dominance of genes related to an adaptive immune response including markers of B cell activity (immunoglobulin-encoding genes, CD79B, BLAME), T cell activity (CD3, TCR α, -β, -δ, LCK), lymph node development (DOCK2, SDF1, CXCR4, MIP1α, TNFSF13b, IP-10, L-selectin and CD8) and ongoing immune response (HLA class II encoding genes, *IFI30*, *IRF1*, *IL2R*, *IL7R*, *STAT1*, *SSI3*). This group consisted of another subgroup expressing higher levels of genes related to the classical pathway of complement activation (C1q, C2, C1R). The second major group expressed genes suggestive for fibroblast dedifferentiation (several types of collagen). Except from tissues with high immune activity, all samples were found to express genes associated with tissue remodeling (Wnt5a, SFRP2, DKK3). Comparing an extended number of synovium samples from RA patients with those of osteoarthritis (OA) patients [32], the same group found evidence for differences in the activation of the STAT1 pathway [33]. These were related to the group of samples with increased immune activity, while the other group of samples with low inflammatory gene expression revealed more activity in tissue remodelling similar to OA samples.

Despite this reported similarity of some RA samples to OA samples, another study aimed at diagnostic patterns [34] demonstrated that 48 genes differentially expressed between RA and OA patients were effective to correctly classify independent samples of synovial tissue for the diagnosis of RA and OA. Concerning the role of tissue homeostasis, Bramlage et al. [35] investigated the contribution of potentially regenerative factors such as bone morphogenetic factors. They showed that BMP4 and BMP5 are decreased in RA compared to OA and normal synovial tissue and the histological distribution of BMP-expressing cells is different in RA.

Comparable to peripheral blood, it has been speculated that the synovial expression pattern of RA samples relates to a possible role of IFN in RA [32, 34]. However, when Nzeusseu Toukap et al. [36] investigated expression profiles of RA, OA and SLE synovial samples, only SLE samples revealed a high IFN-response pattern, whereas RA samples were much lower and more comparable to OA samples. Type I IFN-inducible genes were also found increased in microdissected glomerula of renal biopsies from patients with SLE [37]. Thus, the IFN signature seems to be a special characteristic of SLE patients.

All investigations of inflamed tissue samples reported an over-expression of genes that were typical of either T cells, B cells and/or macrophages. This indicates that infiltration-related expression patterns were identified. However, comparing the expression data from inflamed synovium [38, 39] or inflamed muscle [40, 41] with purified cell types of peripheral blood also indicates that true activation of genes is much higher in inflamed tissue than in peripheral blood samples (see also Tab. 1). Therefore, investigation of purified cells from the inflamed tissue site is also essential.

Earlier investigations by subtractive hybridisation using cultured synovial fibroblasts from RA and OA patients [42] have identified the chemokine SDF1 and several genes capable of mediating synoviocyte-leukocyte interactions, including vascular cell adhesion molecule 1 and Mac-2 binding protein. Extracellular matrix components (lumican, biglycan, and insulin-like growth factor binding protein 5) suggested distinct stromal-synoviocyte interactions. Furthermore, interferon-inducible genes were also found, emphasizing the presence of activation-like features in RA. Kasperkovitz et al. [43] investigated genome-wide microarray expression profiles of cultured synovial fibroblasts of 19 RA patients. They separated synovial cell lines from the different patients into two characteristic groups. One was found to express a myofibroblastoid expression pattern dominated by genes involved in TGFβ/activin A signalling/response pattern (ACTA2, COL3A1, COL4A1/2, SPARC, CYR61, CALD1, IER3, OSF-2, SERPINE1, SERPINH2 and TAGLN). Another group demonstrated an increased expression of IGF2 and IGFBP5, which may indicate involvement of autocrine signalling in the proliferation and survival of synovial fibroblasts. This second cluster also contained over-expression of genes related to complement activation, oxidative stress, and tumor-like features with the expression of different oncogenes (*MAF*, *NBL1*, *MYB*, *TEM8* and *RAB31*). The dominance and importance of TGFβ stimulation was also confirmed by Pohlers et al. [44] when comparing RA with OA synovial fibroblasts.

Further analysis of infiltrating immune cells has been difficult up to now because appropriate technologies to separate individual immune cell types without artificial *in vitro* activation are not yet sufficiently developed. Thus, candidates identified from whole tissue expression profiling are currently investigated stepwise. Timmer et al. [45] analyzed synovial tissues with ectopic lymphoid structure development by selecting for CD21L positive tissue samples. In these histomorphologically confirmed tissues with lymphoid follicles, there was an increase of LTα, LTβ, CXCL13 with its receptor CXCR5, and CCL19 and CCL21 with receptor CCR7, all known to be involved in lymphocyte trafficking to secondary lymphoid organs [46]. Multiclass comparison of expression data between samples with lymphoid structures, samples containing only lymphocyte aggregates and samples with diffuse infiltration revealed a dominance of JAK/STAT pathway-associated genes as well as IL7R and IL2Rγ. The importance of IL7 was histologically confirmed by immunostaining with different typical patterns of IL7 distribution for each type of infiltration. Only lymphoid structures revealed an extracellular staining localized in a circular structure around the B cell follicle.

Functional studies on mechanisms of inflammation and destruction

The synovial fibroblast (SFbl) is one of the most prominent cell types in arthritis pathology [47]. With its different roles in pro-inflammatory mechanisms but also in wound healing as a regenerative potential, the influence of inflammatory triggers and therapeutics is of major interest. Pierer et al. [48] investigated the role of TLR in RA and stimulated RA SFbl with the TLR-2 ligand bacterial peptidoglycan. By microarray analysis they identified upregulation of a wide panel of chemokines and found GCP-2 and MCP-2, as new candidates, expressed in synovial tissues only in RA but not in OA. Ogura et al. [49] stimulated SFbl with IL-1 and found 5 chemokines among the 10 most upregulated genes, headed by CCL20. Similar to these studies, investigations by Santangelo et al. [50] in horse SFbl demonstrated that LPS also induced a broad panel of chemokines as well as cytokines including IL-1, IL-6 and TNFα.

While these studies underline the pro-inflammatory potential of SFbl challenged with TLR ligands or cytokines, Jang et al. [51] and Devauchelle et al. [52] identified, by microarray analysis, genes involved in synovial fibroblast proliferation and survival. mlN51 was found to be upregulated in hyperactive RA SFbl and downregulated in growth-retarded SFbl. GM-CSF and synovial fluid reversed this process, and inhibition of GM-CSF as well as knock-down of mlN51 by siRNA blocked

GM-CSF/synovial fluid mediated proliferation. An important regulator of survival of SFbl seems to be clusterin (CLU). Decreased expression of the intracellular isoform was found in RA compared to OA and normal synovium. While knockdown of CLU with siRNA promoted IL-6 and IL-8 production, transgenic over-expression in RA SFbl of this interactor with phosphorylated IκBα induced apoptosis.

With cartilage destruction as the central pathology of arthritis in mind, we used a previously established model of cartilage tissue engineering to investigate the influence of secreted factors from RA SFbl on cartilage [53]. RA SFbl supernatant contained IL-6, CCL2, CXCL1–3 and CXCL8 as defined by protein array analysis. This was used to stimulate *in vitro* engineered cartilage. Within 48 h the chondrocyte expression profile changed to an inflammatory and destructive pattern, including upregulation of adenosine A2A receptor, cyclooxygenase-2, genes of the NF-κB signalling pathway (toll-like receptor 2, spermine synthase, receptor-interacting serine-threonine kinase 2), cytokines, chemokines and receptors (CXCL1–3, CXCL8, CCL20, CXCR4, IL-1β, IL-6). Furthermore, expression of genes involved in cartilage degradation (MMP-10, MMP-12) were increased, whereas matrix synthesis was suppressed (COMP, chondroitin sulfate proteoglycan 2).

Influence of therapy on gene expression

Different therapeutic concepts have been established to treat rheumatic diseases. None of these is able to cure these diseases and response cannot be predicted for an individual patient. The new biologics target highly specific individual mediators of inflammation. However, not all patients with the same diagnosis seem to benefit in the same way from these expensive drugs. Thus, the central aim is to identify biomarkers that can either indicate response early after starting treatment, or that may even predict response prior to initiation of therapy. Expression profiling of blood cells has been applied to patients with RA prior to anti-TNF therapy by Lequerré et al. [54]. Their analysis revealed a set of genes that was differentially expressed prior to treatment and identified, with up to 90% sensitivity and 70% specificity, responders to infliximab. Interestingly, almost all of the marker genes revealed an increase in responders and a decrease in non-responders after three months of therapy. In an own study by Stuhlmüller et al. [55] gene expression in negatively sorted monocytes was profiled prior to and after anti-TNF treatment. A pattern of marker genes was established that separated healthy controls from RA patients and that distinguished responders from non-responders by grouping non-responders in the cluster of healthy controls. When comparing both groups prior to therapy, integrin alpha-X (CD11c) and two unknown genes (*GS3786*; *FLJ10134*) were differentially expressed. In this context CD11c appears of great interest, since it is expressed on the surface of human monocytes and other cells of the myelomonocytic lineage, and has known functions in inflammatory reactions. Validation by TaqMan real time

PCR in a subsequent study of 12 patients revealed a high correlation with ACR response (R = 0.85). Interestingly, Balanescu et al. [56] found a reduction of CD11c expression of 25% on peripheral blood dendritic cells after 24 h and of 60% after six months of treatment with anti-TNF. In comparison to this promising marker, synoviolin [57] was reported as another predictive marker, however with a false positive (FP) rate of ≈65% for 100% true positives (TP) and 30% FP for 87% TP.

Analysis of synovial tissue biopsies prior to and after anti-TNF treatment was performed by Lindberg et al. [58]. Successful treatment revealed a substantial decrease of several chemokines including CCL3, CCL19, CXCL1, CXCL3, CXCL9 and CXCL10 as well as of TNFα. In particular, patients with high levels of CXCL1, MMP3 and TNFα seemed to be good responders.

To investigate the influence of therapy in a more systematic way, in a first experimental approach we started to incubate SFbl from RA patients and normal joints for 48 h with three different types of drugs, the NSAID diclofenac, the DMARD methotrexate (MTX) and the glucocorticoid prednisolone [10]. While diclofenac did not induce a relevant change in the expression pattern, MTX and prednisolone had different effects. MTX reverted the RA-related expression profile of genes associated with growth and apoptosis, including insulin-like growth factor binding protein 3, retinoic acid induced 3, caveolin 2, and the cell adhesion molecule integrin alpha 6. Prednisolone effectively reverted expression of genes related to inflammation (IL-1β, IL-8). Interestingly, MTX did not reduce but even increased some of the pro-inflammatory cytokines during short term incubation, which may reflect the clinical situation of side effects within the first 1-2 days after administration of this drug.

Future perspectives

Expression profiling in rheumatic diseases is an effective technology to generate an overview of the dominant molecular pathomechanisms. However, current concepts to analyze whole blood and con[illegible]mples is hampered by the fact that the influence of differential cell[illegible]on cannot be sufficiently quantified. Although bioinformatic concep[illegible]oping, reference signatures of highly purified cell types are indispens[illegible] important is the generation of defined stimulation signatures from the [illegible]l types. For example, the cell type specific signature for granulopoies[illegible]okine-response signature for IFN were important in identifying such p[illegible]pression profiles from SLE. In a similar way, signatures of other cell t[illegible]uli are needed to interpret the profiles generated by different diseases[illegible] the development of appropriate bioinformatic tools and databases i[illegible]which should enable the instantaneous retrieval of cell type and stim[illegible]marker genes. Furthermore, comparative analyses between different ty[illegible]es within one disease, as well as between the same type of samples but from different diseases, is essential to assess the importance of a particular stimulus for a new condition. Aiming at these important aspects, we have generated the database, SiPaGene, with constantly increasing numbers of arrays, comparative analyses, retrieval options, and marker sets for functional interpretation. Nevertheless, only limited sets of systematically generated signatures are currently available and systematic extension of these collections will be important work for the future.

References

1 Bang H, Egerer K, Gauliard A, Luthke K, Rudolph PE, Fredenhagen G et al (2007) Mutation and citrullination modifies vimentin to a novel autoantigen for rheumatoid arthritis. *Arthritis Rheum* 56(8): 2503–11

2 Mathsson L, Mullazehi M, Wick MC, Sjoberg O, van Vollenhoven R, Klareskog L et al (2008) Antibodies against citrullinated vimentin in rheumatoid arthritis: Higher sensitivity and extended prognostic value concerning future radiographic progression as compared with antibodies against cyclic citrullinated peptides. *Arthritis Rheum* 58(1): 36–45

3 Riemekasten G, Hahn BH (2005) Key autoantigens in SLE. *Rheumatology (Oxford)* 44(8): 975–82

4 Pfueller B, Wolbart K, Bruns A, Burmester GR, Hiepe F (2001) Successful treatment of patients with systemic lupus erythematosus by immunoadsorption with a C1q column: a pilot study. *Arthritis Rheum* 44(8): 1962–3

5 Jonsdottir T, Gunnarsson I, Risselada A, Welin Henriksson E, Klareskog L, van Vollenhoven RF (2008) Treatment of refractory SLE with rituximab plus cyclophosphamide: clinical effects, serological changes, and predictors of response. *Ann Rheum Dis* 67(3): 330–334

6 Vallerskog T, Gunnarsson I, Widhe M, Risselada A, Klareskog L, van Vollenhoven R et al (2007) Treatment with rituximab affects both the cellular and the humoral arm of the immune system in patients with SLE. *Clin Immunol* 122(1): 62–74

7 Elliott MJ, Maini RN, Feldmann M, Long-Fox A, Charles P, Katsikis P et al (1993) Treatment of rheumatoid arthritis with chimeric monoclonal antibodies to tumor necrosis factor alpha. *Arthritis Rheum* 36(12): 1681–90

8 Campion GV, Lebsack ME, Lookabaugh J, Gordon G, Catalano M (1996) Dose-range and dose-frequency study of recombinant human interleukin-1 receptor antagonist in patients with rheumatoid arthritis. The IL-1Ra Arthritis Study Group. *Arthritis Rheum* 39(7): 1092–101

9 Wendling D, Racadot E, Wijdenes J (1993) Treatment of severe rheumatoid arthritis by anti-interleukin 6 monoclonal antibody. *J Rheumatol* 20(2): 259–62

10 Häupl T, Yahyawi M, Lubke C, Ringe J, Rohrlach T, Burmester GR et al (2007) Gene

expression profiling of rheumatoid arthritis synovial cells treated with antirheumatic drugs. *J Biomol Screen* 12(3): 328–40

11 Baechler EC, Batliwalla FM, Karypis G, Gaffney PM, Moser K, Ortmann WA et al (2004) Expression levels for many genes in human peripheral blood cells are highly sensitive to *ex vivo* incubation. *Genes Immun* 5(5): 347–53

12 Mahr S, Burmester GR, Hilke D, Gobel U, Grützkau A, Häupl T et al (2006) Cis- and trans-acting gene regulation is associated with osteoarthritis. *Am J Hum Genet* 78(5): 793–803

13 Grützkau A, Grün JR, Häupl T, Burmester GR, Radbruch A (2007) Gene expression in inflammatory rheumatic diseases. *Dtsch Med Wochenschr* 132(37): 1888–91

14 Baechler EC, Batliwalla FM, Karypis G, Gaffney PM, Ortmann WA, Espe KJ et al (2003) Interferon-inducible gene expression signature in peripheral blood cells of patients with severe lupus. *Proc Natl Acad Sci USA* 100(5): 2610–5

15 Han GM, Chen SL, Shen N, Ye S, Bao CD, Gu YY (2003) Analysis of gene expression profiles in human systemic lupus erythematosus using oligonucleotide microarray. *Genes Immun* 4(3): 177–86

16 Bennett L, Palucka AK, Arce E, Cantrell V, Borvak J, Banchereau J et al (2003) Interferon and granulopoiesis signatures in systemic lupus erythematosus blood. *J Exp Med* 197(6): 711–23

17 Rus V, Atamas SP, Shustova V, Luzina IG, Selaru F, Magder LS et al (2002) Expression of cytokine- and chemokine-related genes in peripheral blood mononuclear cells from lupus patients by cDNA array. *Clin Immunol* 102(3): 283–90

18 Gu J, Marker-Hermann E, Baeten D, Tsai WC, Gladman D, Xiong M et al (2002) A 588–gene microarray analysis of the peripheral blood mononuclear cells of spondyloarthropathy patients. *Rheumatology (Oxford)* 41(7): 759–66

19 Bovin LF, Rieneck K, Workman C, Nielsen H, Sorensen SF, Skjodt H et al (2004) Blood cell gene expression profiling in rheumatoid arthritis. Discriminative genes and effect of rheumatoid factor. *Immunol Lett* 93(2–3): 217–26

20 Batliwalla FM, Baechler EC, Xiao X, Li W, Balasubramanian S, Khalili H et al (2005) Peripheral blood gene expression profiling in rheumatoid arthritis. *Genes Immun* 6(5): 388–97

21 Begovich AB, Carlton VE, Honigberg LA, Schrodi SJ, Chokkalingam AP, Alexander HC et al (2004) A missense single-nucleotide polymorphism in a gene encoding a protein tyrosine phosphatase (PTPN22) is associated with rheumatoid arthritis. *Am J Hum Genet* 75(2): 330–7

22 Olsen N, Sokka T, Seehorn CL, Kraft B, Maas K, Moore J et al (2004) A gene expression signature for recent onset rheumatoid arthritis in peripheral blood mononuclear cells. *Ann Rheum Dis* 63(11): 1387–92

23 van der Pouw Kraan TC, Wijbrandts CA, van Baarsen LG, Voskuyl AE, Rustenburg F, Baggen JM et al (2007) Rheumatoid arthritis subtypes identified by genomic profiling of peripheral blood cells: assignment of a type I interferon signature in a subpopulation of patients. *Ann Rheum Dis* 66(8): 1008–14

24 van der Pouw Kraan TC, van Baarsen LG, Wijbrandts CA, Voskuyl AE, Rustenburg F, Baggen JM et al (2008) Expression of a pathogen-response program in peripheral blood cells defines a subgroup of Rheumatoid Arthritis patients. *Genes Immun* 9(1): 16–22
25 Lyons PA (2002) Gene-expression profiling and the genetic dissection of complex disease. *Curr Opin Immunol* 14(5): 627–30
26 Baron U, Floess S, Wieczorek G, Baumann K, Grützkau A, Dong J et al (2007) DNA demethylation in the human FOXP3 locus discriminates regulatory T cells from activated FOXP3(+) conventional T cells. *Eur J Immunol* ;37(9): 2378–89
27 Lyons PA, Koukoulaki M, Hatton A, Doggett K, Woffendin HB, Chaudhry AN et al (2007) Microarray analysis of human leucocyte subsets: the advantages of positive selection and rapid purification. *BMC Genomics* 8: 64
28 Stuhlmüller B, Ungethum U, Scholze S, Martinez L, Backhaus M, Kraetsch HG et al (2000) Identification of known and novel genes in activated monocytes from patients with rheumatoid arthritis. *Arthritis Rheum* 43(4): 775–90
29 Biesen R, Demir C, Barkhudarova F, Grün JR, Steinbrich-Zöllner M, Backhaus M et al (2008) Siglec1 is a potential biomarker for monitoring disease activity and success of therapy in inflammatory and resident monocyte subsets of SLE patients. *Arthritis Rheum* 58(4): 1136–1145
30 Lindberg J, af Klint E, Ulfgren AK, Stark A, Andersson T, Nilsson P et al (2006) Variability in synovial inflammation in rheumatoid arthritis investigated by microarray technology. *Arthritis Res Ther* 8(2): R47
31 van der Pouw Kraan TC, van Gaalen FA, Huizinga TW, Pieterman E, Breedveld FC, Verweij CL (2003) Discovery of distinctive gene expression profiles in rheumatoid synovium using cDNA microarray technology: evidence for the existence of multiple pathways of tissue destruction and repair. *Genes Immun* 4(3): 187–96
32 van der Pouw Kraan TC, van Gaalen FA, Kasperkovitz PV, Verbeet NL, Smeets TJ, Kraan MC et al (2003) Rheumatoid arthritis is a heterogeneous disease: evidence for differences in the activation of the STAT-1 pathway between rheumatoid tissues. *Arthritis Rheum* 48(8): 2132–45
33 Kasperkovitz PV, Verbeet NL, Smeets TJ, van Rietschoten JG, Kraan MC, van der Pouw Kraan TC et al (2004) Activation of the STAT1 pathway in rheumatoid arthritis. *Ann Rheum Dis* 63(3): 233–9
34 Devauchelle V, Marion S, Cagnard N, Mistou S, Falgarone G, Breban M et al (2004) DNA microarray allows molecular profiling of rheumatoid arthritis and identification of pathophysiological targets. *Genes Immun* 5(8): 597–608
35 Bramlage CP, Häupl T, Kaps C, Ungethum U, Krenn V, Pruss A et al (2006) Decrease in expression of bone morphogenetic proteins 4 and 5 in synovial tissue of patients with osteoarthritis and rheumatoid arthritis. *Arthritis Res Ther* 8(3): R58
36 Nzeusseu Toukap A, Galant C, Theate I, Maudoux AL, Lories RJ, Houssiau FA et al (2007) Identification of distinct gene expression profiles in the synovium of patients with systemic lupus erythematosus. *Arthritis Rheum* 56(5): 1579–88
37 Peterson KS, Huang JF, Zhu J, D'Agati V, Liu X, Miller N et al (2004) Characterization

of heterogeneity in the molecular pathogenesis of lupus nephritis from transcriptional profiles of laser-captured glomeruli. *J Clin Invest* 113(12): 1722–33

38 Häupl T, Grützkau A, Grün J, Kinne R, Berek C, Stuhlmüller B et al (2005) Dominant role of B-cells and monocytes/macrophages in rheumatoid arthritis based on synovial expression profiles. *Arthritis Rheum* 52(#1084)

39 Häupl T, Grützkau A, Grün J, Baumgrass R, Janitz M, Stuhlmüller B et al (2004) Gene expression profiling in rheumatoid arthritis: unraveling complexity by comparison with profiles from purified leukocytes. *Ann Rheum Dis* 63 (Suppl. 1): 54

40 Nagaraju K, Casciola-Rosen L, Lundberg I, Rawat R, Cutting S, Thapliyal R et al (2005) Activation of the endoplasmic reticulum stress response in autoimmune myositis: potential role in muscle fiber damage and dysfunction. *Arthritis Rheum* 52(6): 1824–35

41 Nagaraju K, Rider LG, Fan C, Chen YW, Mitsak M, Rawat R et al (2006) Endothelial cell activation and neovascularization are prominent in dermatomyositis. *J Autoimmune Dis* 3: 2

42 Seki T, Selby J, Häupl T, Winchester R (1998) Use of differential subtraction method to identify genes that characterize the phenotype of cultured rheumatoid arthritis synoviocytes. *Arthritis Rheum* 41(8): 1356–64

43 Kasperkovitz PV, Timmer TC, Smeets TJ, Verbeet NL, Tak PP, van Baarsen LG et al (2005) Fibroblast-like synoviocytes derived from patients with rheumatoid arthritis show the imprint of synovial tissue heterogeneity: evidence of a link between an increased myofibroblast-like phenotype and high-inflammation synovitis. *Arthritis Rheum* 52(2): 430–41

44 Pohlers D, Beyer A, Koczan D, Wilhelm T, Thiesen HJ, Kinne RW (2007) Constitutive upregulation of the transforming growth factor-beta pathway in rheumatoid arthritis synovial fibroblasts. *Arthritis Res Ther* 9(3): R59

45 Timmer TC, Baltus B, Vondenhoff M, Huizinga TW, Tak PP, Verweij CL et al (2007) Inflammation and ectopic lymphoid structures in rheumatoid arthritis synovial tissues dissected by genomics technology: identification of the interleukin-7 signaling pathway in tissues with lymphoid neogenesis. *Arthritis Rheum* 56(8): 2492–502

46 Weyand CM, Goronzy JJ (2003) Ectopic germinal center formation in rheumatoid synovitis. *Ann NY Acad Sci* 987: 140–9

47 Muller-Ladner U, Ospelt C, Gay S, Distler O, Pap T (2007) Cells of the synovium in rheumatoid arthritis. Synovial fibroblasts. *Arthritis Res Ther* 9(6): 223

48 Pierer M, Rethage J, Seibl R, Lauener R, Brentano F, Wagner U et al (2004) Chemokine secretion of rheumatoid arthritis synovial fibroblasts stimulated by Toll-like receptor 2 ligands. *J Immunol* 172(2): 1256–65

49 Ogura N, Akutsu M, Tobe M, Sakamaki H, Abiko Y, Kondoh T (2007) Microarray analysis of IL-1beta-stimulated chemokine genes in synovial fibroblasts from human TMJ. *J Oral Pathol Med* 36(4): 223–8

50 Santangelo KS, Johnson AL, Ruppert AS, Bertone AL (2007) Effects of hyaluronan

treatment on lipopolysaccharide-challenged fibroblast-like synovial cells. *Arthritis Res Ther* 9(1): R1

51 Jang J, Lim DS, Choi YE, Jeong Y, Yoo SA, Kim WU et al (2006) mlN51 and GM-CSF involvement in the proliferation of fibroblast-like synoviocytes in the pathogenesis of rheumatoid arthritis. *Arthritis Res Ther* 8(6): R170

52 Devauchelle V, Essabbani A, De Pinieux G, Germain S, Tourneur L, Mistou S et al (2006) Characterization and functional consequences of underexpression of clusterin in rheumatoid arthritis. *J Immunol* 177(9): 6471–9

53 Andreas K, Lubke C, Häupl T, Dehne T, Morawietz L, Ringe J et al (2008) Key regulatory molecules of cartilage destruction in rheumatoid arthritis: an *in vitro* study. *Arthritis Res Ther* 10(1): R9

54 Lequerré T, Gauthier-Jauneau AC, Bansard C, Derambure C, Hiron M, Vittecoq O et al (2006) Gene profiling in white blood cells predicts infliximab responsiveness in rheumatoid arthritis. *Arthritis Res Ther* 8(4): R105

55 Stuhlmüller B, Häupl T, Tandon N, Hernandez M, Hultschig C, Kuban RJ et al (2005) Microarray analysis for molecular characterization of disease activity and measuring outcomes of anti-tumour necrosis factor therapy in rheumatoid arthritis. *Arthritis Res Ther* 7 (Suppl 1): P159

56 Balanescu A, Radu E, Nat R, Regalia T, Bojinca V, Ionescu R et al (2005) Early and late effect of infliximab on circulating dendritic cells phenotype in rheumatoid arthritis patients. *Int J Clin Pharmacol Res* 25(1): 9–18

57 Toh ml, Marotte H, Blond JL, Jhumka U, Eljaafari A, Mougin B et al (2006) Overexpression of synoviolin in peripheral blood and synoviocytes from rheumatoid arthritis patients and continued elevation in nonresponders to infliximab treatment. *Arthritis Rheum* 54(7): 2109–18

58 Lindberg J, af Klint E, Catrina AI, Nilsson P, Klareskog L, Ulfgren AK et al (2006) Effect of infliximab on mRNA expression profiles in synovial tissue of rheumatoid arthritis patients. *Arthritis Res Ther* 8(6): R179

miRNA patterns in hematopoietic malignancies

Astrid Novosel and Arndt Borkhardt

Universitätsklinikum Düsseldorf, Klinik für Kinder-Onkologie, -Hämatologie und Klinische Immunologie, Moorenstr.5, 40225 Düsseldorf, Germany

Abstract

MicroRNAs (miRNAs) are small, highly conserved RNA molecules that play key regulatory roles in a diverse range of pathways including control of haematopoiesis, developmental timing, cell differentiation, apoptosis, cell proliferation, and organ development. Bioinformatic data indicate that each miRNA can control hundreds of gene targets, potentially influencing almost every genetic pathway.

The potential importance of microRNAs in cancer is implied by the finding that the majority of human microRNAs are located at cancer-associated genomic regions and there is rapidly accumulating evidence to suggest that dysfunctional expression of microRNAs is a common feature of malignancy. This exciting topic of subtle interactions and regulations within transcription and translation will remain one of the most investigated fields, especially with regard to possibilities for new medications. Different techniques and technologies are currently used to identify and approve miRNA sequences and functions. Among them, miRNA microarray technology – with several updates to be implemented – might then become the preferred platform for whole genome miRNA expression analysis, attaining value in everyday clinical use.

MicroRNAs and their impact on the hematopoietic system and malignancies

MicroRNAs (miRNAs) are small, often phylogenetically highly conserved, non-protein-coding RNAs that form an extensive class of RNA molecules regulating gene expression at post-transcriptional level.

Since the discovery of the first miRNA in the worm *C. elegans* in 1993 [1], much progress has been made in dissecting the biogenesis and functions of miRNAs. The number of microRNAs is hard to estimate at this point – it differs from predicted thousands of miRNAs to the recently published comprehensive cloning and sequencing effort of 172 human, 64 mouse and 16 rat small RNA libraries extracted from major organs and cell types [2]. The emerging evidence suggests that miRNAs are broadly implicated in gene regulation.

A single miRNA can target several hundred genes – it is currently believed that between 10%-30% of all human genes are targets for miRNA regulation [3, 4].

Microarrays in Inflammation, edited by Andreas Bosio and Bernhard Gerstmayer

Definition

MiRNAs are found in multi-cellular plant and animal species, but not in unicellular organisms such as yeast. By definition, miRNAs correspond to cloned, small RNAs that are processed from stem-loop precursors transcribed from genes other than those they regulate [5]. Endogenous, small RNAs that do not meet these criteria, and are known collectively as small-interfering (si)RNAs, derive from long double-stranded RNA precursors that, in some species, may be transcribed from repeat elements, heterochromatin or transposons [6]. Their sequences usually perfectly match those of their RNA targets, while miRNAs recognize target sites – most commonly found in the 3'untranslated regions (UTRs) of cognate mRNA – through imperfect base-pairing, with one or more mismatches in sequence complimentarity. The two mentioned classes of short regulatory RNAs differ mainly in their origins and not in their functions [7].

Impact of miRNAs

The potential and biological impact of microRNA effects has been nicely demonstrated by Lim et al. [8] who showed that delivering miR-124 causes the expression profile in human cell lines to shift towards that of brain, the organ in which miR-124 is preferentially expressed, whereas delivering miR-1 shifts the profile towards that of muscle, where miR-1 is preferentially expressed. This suggested that tissue-specific miRNAs seem to down regulate a far greater number of targets than previously appreciated, thereby helping to define tissue-specific gene expression in humans, generating and maintaining tissue identity.

The enormous importance of microRNA in various key regulatory roles in all kind of pathways – such as developmental timing, cell cycle, cell differentiation and proliferation, apoptosis, organ development, haematopoiesis, cancer, infectious disease, genetic disorders, heart disease – has been shown and is still under investigation [1, 9–13]. The estimated number of miRNA genes is growing steadily, reaching tens of thousands – found by random sequencing, by bioinformatic predictions or comparative genomics, and relying on the conservation of miRNAs between species [14].

How to find miRNAs – and their targets

There are several approaches to finding new miRNAs – most common is the matching of bioinformatic data (see below) and cloning. But then, those miRNAs whose expression is restricted to less-abundant cell types or specific events in cell metabolism or development could be missed in cloning efforts. Another (computational) approach is to search for homologues of known miRNA genes [15,16] to search in the vicinity of known miRNA genes for other stem loops that might harbour miRNA genes of a genomic cluster [17, 18] or to identify conserved genomic segments that are outside of predicted protein-coding regions and potentially could form stem loops – e.g. MiRscan or miRseeker [19–21].

Cloning individual genes to determine whether they code for miRNAs is difficult and the bioinformatics approach requires experimental verification. One of the huge databases where the scientific community deposits newly-found miRNAs is miRBase: http://microrna.sanger.ac.uk/.This database provides a searchable online repository for published microRNA sequences and associated annotation, functionality previously provided by the microRNA Registry. miRBase also contains predicted miRNA target genes in miRBase Targets, and provides a gene naming and nomenclature function in the miRBase Registry. The latest release (August 2007) of the database contains 5071 entries, thereof 533 human miRNAs.

Cummins et al. (2006) have developed an experimental approach called miRNA serial analysis of gene expression (miRAGE) to search for other miRNAs. Their study shows that previous predictions of the total number of human miRNAs were too low. Basically, this method involves isolating thousands of tiny RNA molecules from cells. These RNA molecules are reverse-transcribed into cDNA, then linked together into larger chains and sequenced. The researchers then used bioinformatics to analyze the DNA sequences (for similarity to miRNA genes in other species, ability to form a hairpin). The sequence analysis of nearly 274,000 small RNA tags allowed them to identify 200 already known miRNAs and 133 novel miRNA candidates [19, 20, 22]. With the miRAGE method they increased the number of experimentally verified microRNAs by almost 50 percent. There are estimated to be more than 1000 miRNAs in the mammalian genome, regulating around 30% of the protein coding genes [4, 23–25].

To facilitate the investigation of new miRNAs, Calin et al. showed that miR genes are frequently located at fragile sites, as well as in minimal regions of loss of heterozygosity, minimal regions of amplification (minimal amplicons), or common breakpoint regions. Additionally, by Northern blotting, this group has shown that several miRs located in deleted regions have low levels of expression in cancer samples [26].

Huppi et al. have extended this observation to a large-scale study of overlapping retroviral integration sites (RIS)/translocation (Tx) breakpoints (~345) and the genomic location of miRs (~319) in the mouse. They utilized the Jenkins-Copeland mouse Retrovirus Tagged Cancer Gene Database and the Sanger miRNA registry and found that a significantly large number (47%) of mouse miRs reside in clusters (~22–51), many of which overlap with clusters of RIS. Computer-based interspecies comparisons of human and mouse miR sequence locations and sequences revealed some interesting candidate miRs that may play a role in human disease. This group observed that clusters of RIS were not associated with any known miRs, suggesting potential areas on which to focus in the search for as yet undiscovered miRNAs [27].

Miscellanous platforms have been developed for more or less routine analysis of miRNAs, for example: Northern Blots – with the advantage that new miRNAs were found [28], bead-based hybridization – which features superior specificity [29], *in situ* hybridization – unveils miRNAs in tissue context [30], single molecule detection – permits a quick assay [31], cloning and sequencing – as performed by Landgraf et al. to provide an miRNA expression atlas based on small RNA library sequencing [2,22]. Other methods such as real time PCR have the advantage of comparatively low cost, high throughput, and very good detection of low-abundance species [32]. A PCR based method for higher throughput approaches is e.g. the multiplex qRT-PCR system that reverse transcribes and amplifies up to 330 miRNAs simultaneously with a divided subsequent quantitative PCR reaction [33]. The advantage of this method is that it can be used for samples with extremely low input, such as single embryonic stem cells [34] and that one run can be done very quickly. Other methods, such as RNase protection assays [35–37] or primer extension methods [38] and Invader assays [39] have also been utilized and can be applied to a high-throughput scenario.

miRNA target prediction

As reviewed by Rajewski [20, 40] there are several algorithms to predict miRNA targets and, at present, it is difficult to judge which algorithm (see Tab. 1) is the most reliable or sensible.

The big challenge that the scientific community still faces is to confirm these predictions by experimental data.

MicroRNA biosynthesis

Most miRNA genes come from regions of the genome quite distant from previously annotated genes, implying that they derive from independent transcription units [15], others are in the introns of pre-mRNA, preferentially in the same orientation as the predicted mRNA, suggesting that most of these miRNAs are not transcribed from their own promotors but are processed from the introns [15, 17, 21, 41] (see Fig. 1).

MicroRNA molecules are produced by RNA polymerase II forming long (up to several kb) primary miRNAs – pri-miRNAs – that contain a 5'CAP structure and are polyadenylated at their 3'end. These pri-miRNAs embody one or more stem-loop or hairpin structures of ~70nt that are recognized and cleaved by the nuclear 650kDa microprocessor complex consisting of a ds-RNA specific RNase III endonucleases

Table 1

Precomputed predictions on searchable websites	Organisms	Website
miRNA target predictions at EMBL	flies	http://www.russell.embl-heidelberg.de/miRNAs/
miRanda	flies, vertebrates	http://www.microrna.org//miranda.html
mirBase	vertbrates, insects, nematodes	http://microrna.sanger.ac.uk/targets/v2/
PicTar	vertebrates, flies, nematodes	http://pictar.bio.nyu.edu
TargetScan, TargetScanS	vertebrates	http://genes.mi.edu/targetscan
Ref.27	flies, nematodes	http://tavazoielab.princeton.edu/mirnas/
RNA hybrid	flies	http://www.techfak.uni-bielefeld.de/persons/marc/mirna/targets/drosophila
Tools for locating miRNA targets		
RNAhybrid		http://bibiserv.techfak.uni-bielefeld.de/rnahybrid/welcome.html
DIANA-MicroT		http://diana.pbci.upenn.edu/DIANA-microT
RNA22		http://cbcsrv.watson.ibm.com/rna22.html
Databases for targets with experimental support		
Tarbase		http://www.diana.pcbi.upenn.edu/arbase.html
Argonaute		http://rna.uni-heidelberg.de/apps/zmf/argonaute/interface
miRNAMAP		http://nirnamap.mbc.nctu.edu.tw/

(Rajewsky 2006)

Drosha and the ds-RNA binding protein DGCR8. The resulting ~60–100 nucleotide RNA hairpin intermediate with a two nucleotide 3'overhang is then shuttled by the nuclear export factor Exportin-5 and its cofactor Ran-GTP to the cytoplasm. There, the precursor miRNA (pre-miRNA) hairpin is cleaved by Dicer – a second RNaseIII endonucleases with its ds-RNA binding partner TRBP in humans (or Loqs in flies), near the loop to create a ~20–25 nucleotide imperfect miRNAmiRNA duplex [42, 43]. Then, TRBP recruits the human Argonaute protein (hAgo2) to the Dicer complex,

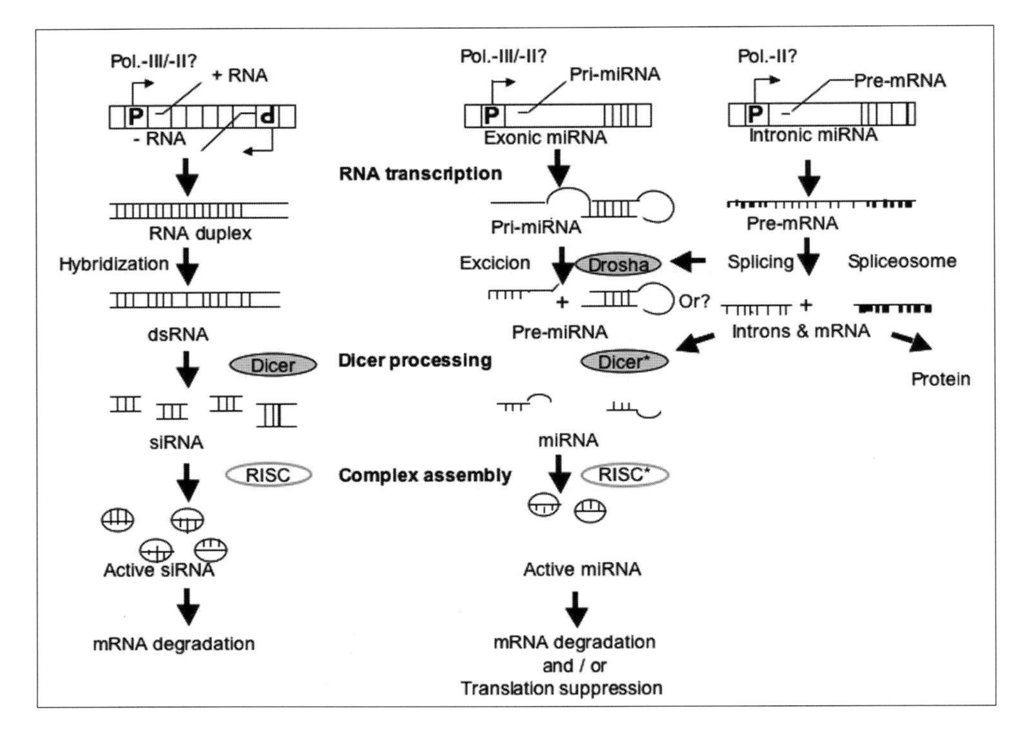

Figure 1.
Comparison of the siRNA, intergenic (exonic) miRNA and intronic miRNA
siRNA is formed by perfectly complementary RNAs and processed into 19–22bp duplexes by the RNase III-familial endonuclease Dicer.
The biogenesis of intergenic (exonic) miRNAs requires a long transcript precursor (pri-miRNA), which is probably generated by Pol-II or Pol-III RNA promoters.
Intronic miRNAs are transcribed by the Pol-II promoters of its encoded genes and coexpressed in the intron regions of the gene transcripts. Lin et al proposed in their model, that after RNA splicing and further processing, the spliced intron may function as a pri-miRNA for intronic miRNA generation. In the nucleus, the pri-miRNA is excised by Drosha RNase to form a hairpin-like pre-miRNA template and then exported to the cytoplasm for further processing by Dicer to form mature miRNA.
The Dicers for siRNA and miRNA pathways are different.
All the described small regulatory RNAs are finally incorporated into an RNA-induced silencing complex (RISC), which contains either the strand of siRNA or the single strand of miRNA. siRNA primarily triggers mRNA degradation, whereas miRNA can induce either mRNA degradation or suppression of protein synthesis depending on the sequence complementarity to the target gene transcripts. (Lin et al 2006)

forming an RNA-induced silencing complex – RISC [44–47]. RISC was identified previously as repressing the translation of mRNA in the presence of small interfer-

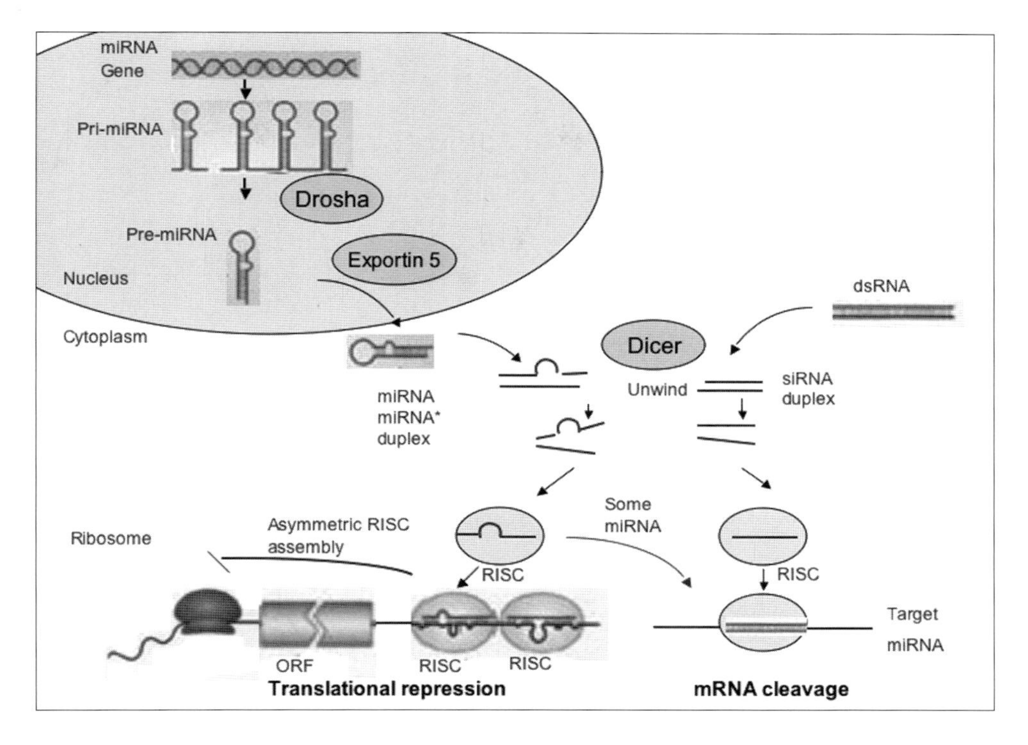

Figure 2.
The RNAi pathway: Long double stranded RNA is cleaved by the RNase III-type endonuclease Dicer. The products are small interfering RNAs (siRNAs) app. 21 nucleotides in length. After incorporation of the siRNA into the RNAi-inducing silencing complex (RISC) the siRNA-molecule becomes unwound in an ATP-dependent manner. The siRNA guides the complex by identifying the target mRNA through base-pairing and directs mRNA cleavage across from the center of the siRNA.
The miRNA pathway: A precursor molecule of app. 70 nucleotides forming a stem loop structure is cleaved by Dicer. The product is a single stranded micro RNA molecule (miRNA) targeting the 3'-UTR of an mRNA. In contrast to siRNAs the miRNA and its target sequence are not completely complementary showing characteristic bulges. The binding of the miRNA to the mRNA leads to the inhibition of translation. (He and Hannon, 2004)

ing RNAs (siRNAs) by an RNA interference (RNAi) mechanism. The essential core components of this complex are members of the Argonaute family which contain two conserved RNA-binding domains: the PAZ domain, which binds to the single-stranded 3'end of the miRNA, and the PIWI domain, which is structurally like ribonuclease-H and which interacts with the miRNA guide strand at its 5'end [48] (see Fig. 2). The result is a decreased level of the protein encoded by the mRNA. A single miRNA can bind multiple genes and have a possible profound effect on cell physiology.

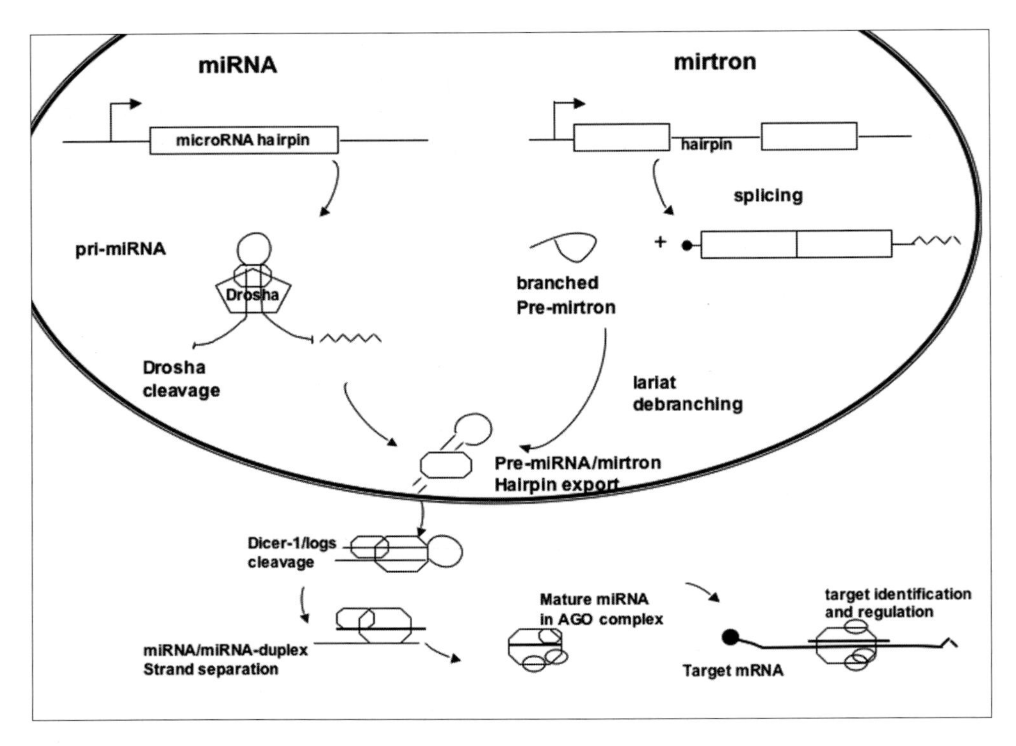

Figure 3.
Model for the convergence of the Mirtron and canonical miRNA pathways
The canonical miRNA pathway initiates with the recognition and cleavage of pri-miRNA transcripts by the Pasha/Drosha complex to yield pre-miRNA hairpins. There exists an alternate pathway in which short introns with hairpin potential are spliced and debranched to yield mirtron hairpins. Both pre-miRNA and mirtron hairpins are exported from the nucleus by Exportin-5 and cleaved by Dicer-1/loqs to generate 22 nt RNA duplexes. One strand, the active miRNA, is transferred to an Ago complex and guides it to repress fully complementary or seed-matched target transcripts. (Okamura et al 2007)

Another recently discovered pathway in Drosophila and C. *elegans* [49, 50], the so-called mirtron pathway, appears to bypass Drosha and Dicer and seems to be essential. As some of these mirtrons have been maintained during evolution with patterns of sequence conservation, this is a hint that this pathway could be an early version in the appearance of miRNA before the Drosha development; it appears to play a crucial role in regulatory functions. In this pathway debranched introns mimic the structural features of pre-miRNA to enter the mi-RNA-processing pathway (Fig. 3). The mirtron hairpins are processed by DBR1- (a RNA lariat debranching enzyme, involved in intron turnover) then slip into the canonical miRNA pathway during hairpin export by exportin-5 and are subsequently processed by Dicer-1/loqs as miRNAs coming from Drosha/ Dicer. In summary, mirtrons seem to produce miRNAs that associate with Ago1, can actively repress minimally paired see targets and display a pattern of divergence on microevolutionary scales that indicates their incorporation into endogenous regulatory networks.

Impact of miRNAs on haematopoietic system and malignancies

Haematopoietic lineage differentiation, the process of continuous development of haematopoietic stem cells into at least eight different blood lineages, is known to be controlled or modulated by complex molecular events that simultaneously regulate the commitment, proliferation, apoptosis, and maturation of haematopoietic stem/progenitor cells. The demonstration that certain miRNAs are differentially expressed in haematopoietic lineages *in vivo* and are able to alter lineage differentiation provides solid evidence that miRNAs represent a class of molecules that regulate mouse haematopoiesis and, more broadly, mammalian development

Chen at al. have shown that there seem to be some blood restricted miRNAs, as miR-181, miR-233 and miR-142. MiR-181 was strongly expressed in thymus (mainly containing T lymphocytes), strongly in brain and lung and detectable in BM and spleen, miR-223 nearly exclusively expressed in bone marrow, and miR-142 (whose gene is at the site of a translocation associated with aggressive B cell leukaemia) was highly expressed in all haematopoietic tissues, but not otherwise.

The differential expression in different haematopoietic lineages of these 3 miRNAs has been demonstrated and suggests an influence in the haematopoietic lineage of these miRNAs on differentiation. As miR-181 has an influence on B as well as on T cells, which are not linked during lineage comittement, this suggests that this miR acts indepently on these two lineages, perhaps through regulation of different target genes. The miRanda algorithm, whose predicted targets overlap minimally with others, hints at the human HOXA5 gene as a candidate miR-181 target. This putative target gene is expressed in haematopoietic cells, where its has previously been implicated in regulating the balance between myeloid and erythroid cell precursors [51] – only an over-expression of miR-181a has no impact on erythroid cell differentiation [13] so the interaction between miR-181 and HOXA5 is still uncertain.

By extensive cloning and bioinformatic efforts, Landgraf et al., in a comparison of miRNA expression in the haematopoietic system with all other organ systems, found that only five miRNAs are highly specific for haematopoietic cells: miR-142, miR-144, miR-150, miR-155 and miR-223 [2].

A majority of human miRNAs are located at cancer-associated genomic or so-called fragile regions and there is growing evidence that dysfunctional expression of miRNAs is a common feature of malignancies [26, 52, 53].

Fulci et al. provide an informative profile of the expression of miRNAs in primary chronic lymphocytic leukemia (CLL) cells using two independent and quantitative methods: with both methods – miRNA cloning and quantitative real-time-polymerase chain reaction (qRT-PCR) of mature miRNAs. They showed that miR-21 and miR-155 are dramatically over-expressed in patients with CLL, although the corresponding genomic loci are not amplified, miR-150 and miR-92 are significantly deregulated in patients with CLL, and miR-15a and miR-16 show decreased levels in about 11% of cases. Additionally, a set of miRNAs whose expression correlates with biologic parameters of prognostic relevance, particularly with the mutational status of the IgV(H) genes, was identified [54]. Studies such as this show that miRNAs have – as also discussed for other malignancies – either a direct or an indirect role [55] and impact multifactorial development in haematopoietic malignancies, and might offer a reasonable therapeutic target in the future.

Another highly specific microRNA, miR-233 in haematopoietic tissue [2], was found – expression level dependent – to be crucial in the differentiation fate of myeloid precursors [54, 56, 57].

The activation of two pathways – of transcriptional regulation by the myeloid lineage-specific transcription factor C/EBPalpha (CCAAT/enhancer-binding protein-alpha) and posttranscriptional regulation by miR-223 – appears to be essential for granulocytic differentiation and clinical response of acute promyelocytic leukaemia (APL) blasts to all-trans retinoic acid (ATRA) [57].

The finding that down regulation of miR-16-1 and miR-15a expression correlates with allelic loss at 13q14 may be of importance to clinical classification of CLL. Patients with a normal karyotype or deletion of 13q14 as the single genetic abnormality have a better prognosis than those with a complex karyotype or frequent deletion of 11q23 or 17p13 [58]. The miRNA expression profile is associated with progression in CLL and can serve as a possible prognostic marker.

Following the initial finding on miR-15 and miR-16 in CLL, miRNA expression deregulation has also been demonstrated in other tumours.

It has been shown that the microRNA miR-155 can be processed from sequences present in BIC RNA, a spliced and polyadenylated but non-protein-coding RNA that accumulates in lymphoma cells. Clinical isolates of several types of B cell lymphomas, including diffuse large B cell lymphoma (DLBCL), have 10- to 30-fold higher copy numbers of miR-155 than do normal circulating B cells. Similarly, the quantities of BIC RNA are elevated in lymphoma cells, but ratios of the amounts of the two RNAs are not constant, suggesting that the level of miR-155 is controlled by transcription and processing. Significantly higher levels of miR-155 are present in DLBCLs, with an activated B cell phenotype, than with the germinal center phenotype. Because patients with activated B cell-type DLBCL have a poorer clinical prognosis, quantification of this microRNA may be diagnostically useful [59] Interestingly, Metzler et al. [60] have demonstrated over-expression of precursor microRNA-155/BIC RNA in children with Burkitt lymphoma. Kluiver et al. have found a lack of BIC and miR-155 expression in this type of lymphoma and have provided evidence for two levels of regulation for mature miR-155 expression in more recent data: one at the transcriptional level involving PKC and NF-kappaB, and one at the processing level. This means that Burkitt lymphoma cells not only express low levels of BIC, but also prevent processing of BIC *via* a mechanism as yet unknown [60–63].

Microarray-based expression studies have revealed alterations in human miRNA expression profiles derived from the miR-17–92 cluster (named „OncomiR-1“) which is located at 13q31.3, a genomic locus that is amplified in cases of diffuse large B cell lymphoma, follicular lymphoma, mantle cell lymphoma, primary cutaneous B cell lymphoma and other tumour types [64]. The transcript of this cluster appears to be the functional precursor of seven microRNAs: miR-17-5p, miR-17-3p, miR-18, miR-19a, miR-20, miR-19b-1 and miR-92-1. Additionally, this cluster is related to the homologous miR-106a–92 cluster on chromosome×and the miR-106b–25 cluster on chromosome 7 [65]. Co-expression of the miR-17–92 cluster acted with c-myc expression to accelerate tumour development in a mouse B-cell lymphoma model [64].

O'Donnell et al. [66] discovered that c-Myc negatively regulates the transcription of miR-17-92 cluster and directly binds the genomic locus encoding these miRNAs. Dysregulated expression or function of c-Myc is one of the most common abnormalities in human malignancy. They also showed that two miRNAs of miR-17-92 cluster (miR-17-5p and miR-20a) target an important proliferative/proapoptotic transcription factor E2F1 [66].

Another miRNA located at a site of rearrangement linked to human leukaemia is miR-142, whose gene is at the breakpoint junction of a t(8;17) translocation, which causes an aggressive B cell leukaemia due to up regulation of a translocated c-MYC gene [16].

Another absorbing insight into the role of miRNAs – in this case miR-125s – was provided by the case report of Sonoki et al. [67]. Showing an unusual IGH rearrangement exhibiting insertion of miR-125b-1 sequence in a patient with B-ALL, they hypothesized that mir-125s could be involved in the timing of tissue development or cell differentiation. With the insertion of miR125b-1 into the IGH allele the expression of the small RNA might be disorganized in precursor B cells and thus lead to aberrant differentiation.

There are many studies coming up soon that will be interesting to discuss and follow in terms of how these cognitions will lead to future medication or, more generally, to a deeper understanding of different diseases. For example, there seems to be evidence that miRNAs are involved in T-ALL. This was investigated by using a miRNA gene chip. Differentially expressed miRNAs were observed in T-ALL vs. the normal (thymocyte) samples. miR-23 emerged as one of the most significantly down regulated miRNA genes in T-ALL, a down regulation that was also validated by Northern blot and both precursor and mature Q-PCR analyses. Mature miR-23 can be derived from two different precursors residing on chromosomes 9q22 (miR-23a) and 19p13 (miR-23b). MiR-23 is also part of a duplicate three miRNA-gene cluster with homologues of miR-24 and miR-27 on both chromosomes. Expression of both miR-23a (9q22) and miR-23b (19p13) were significantly down regulated in T-ALL compared to normal thymocytes. The combined down regulation of related miR genes on 9q22 suggests that down regulation of the entire primary transcript is occurring in T-ALL. However, the observation that related genes on 19p13 are also among the most significantly down regulated miR genes in T-ALL increases the likelihood that one or more genes of this cluster play a role in T-ALL biology. [68].

Another approach is to correlate miRNA expression with global gene expression, which, amongst others, is currently being done by Bullinger et al. [69, 70], supporting a potential role of differential miRNA expression in A ml pathogenesis. Ovcharenko et al. compared three different subpopulations of A ml (standard risk cytogenetics, vs. inv 16 vs. momsomy-7) to normal bone marrow by miRNA microarray analysis. This showed that the median expression of mir-23a, mir-223 and mir-16 was lowest in patients with standard risk cytogenetics, defining mir-23 as a negative prognostic factor, mir-335 was down regulated in momsomy-7 A mls and mir-150 was shown to be highly expressed in monosomy-7 A mls as well as in immature haematopoietic cells ($CD34^+$ stem cells). This and similar studies are likely to provide additional insights into A ml biology, thereby helping to unravel the role of miRNAs in leukemogenesis.

As evidence for deregulated miR expression in haematopoietic malignancies increases daily, this exciting topic of subtle interactions and regulations within transcription and translation will remain one of the most investigated fields, especially with regard to possibilities of new medications.

Future perspectives

Microarray analysis has been used in particular with the ambition of reclassifying carcinomas at the molecular level, to aid in diagnosis/prognosis, and to predict how various types of tumours respond to different therapeutic agents. The massive amount of data – and especially bringing different array data together, e.g. SNP arrays and Gene chip data or Proteomics – is one of the challenges of the near future. Bioinformatics is now on the front line of assistance, creating tools to understand these enormous datasets.

In the future, microarray data could ideally be available for each patient as a bedside test to classify and subclassify different diseases and thus to optimize therapeutic strategies.

References

1 Lee, R.C., R.L. Feinbaum, V. Ambros. 1993. The *C. elegans* heterochronic gene lin-4 encodes small RNAs with antisense complementarity to lin-141. *Cell* 75.843–54

2 Landgraf, P, M. Rusu, R. Sheridan, A. Sewer, N. Iovino, A. Aravin, S. Pfeffer, A. Rice, A.O. Kamphorst, M. Landthaler et al. 2007. A mammalian microRNA expression atlas based on small RNA library sequencing. *Cell* 129.1401–14

3 John, B., A.J. Enright, A. Aravin, T. Tuschl, C. Sander, D.S. Marks. 2004. Human MicroRNA targets. *PLoS Biol* 2.e363

4 Lewis, B.P, C.B. Burge, D.P. Bartel. 2005. Conserved seed pairing, often flanked by adenosines, indicates that thousands of human genes are microRNA targets 1. *Cell* 120.15–20

5 Shivdasani, R.A. 2006. MicroRNAs: regulators of gene expression and cell differentiation. *Blood* 108.3646–53

6 Zamore, P.D, B. Haley. 2005. Ribo-gnome: the big world of small RNAs. *Science* 309.1519–24

7 Ambros, V., B. Bartel, D.P. Bartel, C.B. Burge, J.C. Carrington, X. Chen, G. Dreyfuss, S.R. Eddy, S. Griffiths-Jones, M. Marshall et al. 2003. A uniform system for microRNA annotation. *RNA* 9.277–9

8 Lim, L.P., N.C. Lau, P. Garrett-Engele, A. Grimson, J.M. Schelter, J. Castle, D.P. Bartel, P.S. Linsley, J.M. Johnson. 2005. Microarray analysis shows that some microRNAs downregulate large numbers of target mRNAs. *Nature* 433.769–73

9 Kim, V.N. 2005. MicroRNA biogenesis: coordinated cropping and dicing. *Nat Rev Mol Cell Biol* 6.376–85

10 Lin, S. L, S. J. Chang, and S. Y. Ying. 2006. First *in vivo* evidence of microRNA-induced fragile × mental retardation syndrome. *Mol Psychiatry* 11.616–7

11 van Rooij E., L.B. Sutherland, N. Liu, A.H. Williams, J. McAnally, R.D. Gerard, J.A. Richardson, E.N. Olson. 2006. A signature pattern of stress-responsive microRNAs that can evoke cardiac hypertrophy and heart failure. *Proc Natl Acad Sci USA* 103.18255–60

12 Chen, C.Z., H.F. Lodish. 2005. MicroRNAs as regulators of mammalian hematopoiesis. *Semin Immunol* 17.155–65

13 Chen, C.Z, L. Li, H.F. Lodish, D.P. Bartel. 2004. MicroRNAs modulate hematopoietic lineage differentiation. *Science* 303.83–6

14 Berezikov, E., E. Cuppen, R. H. Plasterk. 2006. Approaches to microRNA discovery. *Nat Genet* 38 Suppl.S2–S7

15 Lagos-Quintana, M., R. Rauhut, W. Lendeckel, T. Tuschl. 2001. Identification of novel genes coding for small expressed RNAs. *Science* 294.853–8
16 Lagos-Quintana, M., R. Rauhut, A. Yalcin, J. Meyer, W. Lendeckel, T. Tuschl. 2002. Identification of tissue-specific microRNAs from mouse. *Curr Biol* 12.735–9
17 Aravin, A.A., M. Lagos-Quintana, A. Yalcin, M. Zavolan, D. Marks, B. Snyder, T. Gaasterland, J. Meyer, T. Tuschl. 2003. The small RNA profile during *Drosophila melanogaster* development. *Dev Cell* 5.337–50
18 Sewer, A., N. Paul, P. Landgraf, A. Aravin, S. Pfeffer, M.J. Brownstein, T. Tuschl, Nimwegen E. van, and M. Zavolan. 2005. Identification of clustered microRNAs using an ab initio prediction method. *BMC Bioinformatics* 6.267
19 Bartel, D.P. 2004. MicroRNAs: genomics, biogenesis, mechanism, and function. *Cell* 116.281–97
20 Lai, E.C., P. Tomancak, R.W. Williams, G.M. Rubin. 2003. Computational identification of Drosophila microRNA genes. *Genome Biol* 4.R42
21 Lim, L.P., M.E. Glasner, S. Yekta, C.B. Burge, D.P. Bartel. 2003. Vertebrate microRNA genes. *Science* 299.1540
22 Cummins, J.M., Y. He, R.J. Leary, R. Pagliarini, L.A. Diaz Jr, T. Sjoblom, O. Barad, Z. Bentwich, A.E. Szafranska, E. Labourier, et al. 2006. The colorectal microRNAome. *Proc Natl Acad Sci USA* 103.3687–92
23 Berezikov, E., V. Guryev, Belt J. van de, E. Wienholds, R. H. Plasterk, E. Cuppen. 2005. Phylogenetic shadowing and computational identification of human microRNA genes. *Cell* 120.21–4
24 Berezikov, E, R.H. Plasterk. 2005. Camels and zebrafish, viruses and cancer: a microRNA update. *Hum Mol Genet* 14 Spec No. 2.R183–R190
25 Lewis, B.P., I.H. Shih, M.W. Jones-Rhoades, D.P. Bartel, C.B. Burge. 2003. Prediction of mammalian microRNA targets. *Cell* 115.787–98
26 Calin, G.A., C. Sevignani, C.D. Dumitru, T. Hyslop, E. Noch, S. Yendamuri, M. Shimizu, S. Rattan, F. Bullrich, M. Negrini, C.M. Croce. 2004. Human microRNA genes are frequently located at fragile sites and genomic regions involved in cancers. *Proc Natl Acad Sci USA* 101.2999–3004
27 Huppi, K., N. Volfovsky, M. Mackiewicz, T. Runfola, T.L. Jones, S.E. Martin, R. Stephens, N.J. Caplen. 2007. MicroRNAs and genomic instability. *Semin Cancer Biol* 17.65–73
28 Lau, N.C., L.P. Lim, E.G. Weinstein, D.P. Bartel. 2001. An abundant class of tiny RNAs with probable regulatory roles in *Caenorhabditis elegans*. *Science* 294.858–62
29 Lu, J, G. Getz, E.A. Miska, E. varez-Saavedra, J. Lamb, D. Peck, A. Sweet-Cordero, B.L. Ebert, R.H. Mak, A.A. Ferrando, J.R. Downing et al. 2005. MicroRNA expression profiles classify human cancers. *Nature* 435.834–8
30 Thompson, R.C., M. Deo, D.L. Turner. 2007. Analysis of microRNA expression by *in situ* hybridization with RNA oligonucleotide probes. *Methods* 43.153–61
31 Neely, L.A., S. Patel, J. Garver, M. Gallo, M. Hackett, S. McLaughlin, M. Nadel, J.

Harris, S. Gullans, J. Rooke. 2006. A single-molecule method for the quantitation of microRNA gene expression. *Nat Methods* 3.41–6

32 Chen, C., D.A. Ridzon, A.J. Broomer, Z. Zhou, D.H. Lee, J.T. Nguyen, M. Barbisin, N.L. Xu, V.R. Mahuvakar, M.R. Andersen, K.Q. Lao, K.J. Livak, K.J. Guegler. 2005. Real-time quantification of microRNAs by stem-loop RT-PCR. *Nucleic Acids Res* 33.e179

33 Lao, K., N.L. Xu, V. Yeung, C. Chen, K.J. Livak, N.A. Straus. 2006. Multiplexing RT-PCR for the detection of multiple miRNA species in small samples. *Biochem Biophys Res Commun* 343.85–9

34 Tang, F., P. Hajkova, S.C. Barton, K. Lao, M.A. Surani. 2006. MicroRNA expression profiling of single whole embryonic stem cells. *Nucleic Acids Res.* 34.e9

35 Hamilton, A.J., D.C. Baulcombe. 1999. A species of small antisense RNA in posttranscriptional gene silencing in plants. *Science* 286.950–2

36 Sijen, T., R. H. Plasterk. 2003. Transposon silencing in the *Caenorhabditis elegans* germ line by natural RNAi. *Nature* 426.310–4

37 Soutschek, J., A. Akinc, B. Bramlage, K. Charisse, R. Constien, M. Donoghue, S. Elbashir, A. Geick, P. Hadwiger, J. Harborth et al. 2004. Therapeutic silencing of an endogenous gene by systemic administration of modified siRNAs. *Nature* 432.173–8

38 Seitz, H., H. Royo, M.L. Bortolin, S.P. Lin, A.C. Ferguson-Smith, J. Cavaille. 2004. A large imprinted microRNA gene cluster at the mouse Dlk1–Gtl2 domain. *Genome Res* 14.1741–8

39 Allawi, H.T., J.E. Dahlberg, S. Olson, E. Lund, M. Olson, W.P. Ma, T. Takova, B.P. Neri, V.I. Lyamichev. 2004. Quantitation of microRNAs using a modified Invader assay. *RNA* 10.1153–61

40 Rajewsky, N. 2006. microRNA target predictions in animals. *Nat Genet* 38 SupplS8–13

41 Lin, S.L., J.D. Miller, S.Y. Ying. 2006. Intronic MicroRNA (miRNA). *J Biomed Biotechnol* 2006.26818

42 Hammond, S.M., E. Bernstein, D. Beach, G. J. Hannon. 2000. An RNA-directed nuclease mediates post-transcriptional gene silencing in Drosophila cells. *Nature* 404.293–6

43 Hutvagner, G., J. McLachlan, A.E. Pasquinelli, E. Balint, T. Tuschl, P.D. Zamore. 2001. A cellular function for the RNA-interference enzyme Dicer in the maturation of the let-7 small temporal RNA. *Science* 293.834–8

44 Lee, Y., C. Ahn, J. Han, H. Choi, J. Kim, J. Yim, J. Lee, P. Provost, O. Radmark, S. Kim, V.N. Kim. 2003. The nuclear RNase III Drosha initiates microRNA processing. *Nature* 425.415–9

45 Tomari, Y., P. D. Zamore. 2005. Perspective: machines for RNAi. *Genes Dev* 19.517–29

46 Chendrimada, T.P., R.I. Gregory, E. Kumaraswamy, J. Norman, N. Cooch, K. Nishikura, R. Shiekhattar. 2005. TRBP recruits the Dicer complex to Ago2 for microRNA processing and gene silencing. *Nature* 436.740–4

47 Gregory, R.I., T.P. Chendrimada, N. Cooch, R. Shiekhattar. 2005. Human RISC couples microRNA biogenesis and posttranscriptional gene silencing. *Cell* 123.631–40

48 Filipowicz, W. 2005. RNAi: the nuts and bolts of the RISC machine. *Cell* 122.17–20

49 Okamura, K., J.W. Hagen, H. Duan, D.M. Tyler, E.C. Lai. 2007. The mirtron pathway generates microRNA-class regulatory RNAs in Drosophila. *Cell* 130.89–100

50 Ruby, J.G., C.H. Jan, D.P. Bartel. 2007. Intronic microRNA precursors that bypass Drosha processing. *Nature* 448.83–6

51 Crooks, G.M., J. Fuller, D. Petersen, P. Izadi, P. Malik, P.K. Pattengale, D.B. Kohn, J.C. Gasson. 1999. Constitutive HOXA5 expression inhibits erythropoiesis and increases myelopoiesis from human hematopoietic progenitors. *Blood* 94.519–28

52 Si, M.L., S. Zhu, H. Wu, Z. Lu, F. Wu, Y.Y. Mo. 2007. miR-21-mediated tumor growth. *Oncogene* 26.2799–803

53 Hammond, S.M. 2007. MicroRNAs as tumor suppressors. *Nat Genet* 39.582–3

54 Fulci, V., S. Chiaretti, M. Goldoni, G. Azzalin, N. Carucci, S. Tavolaro, L. Castellano, A. Magrelli, F. Citarella, M. Messina et al. 2007. Quantitative technologies establish a novel microRNA profile of chronic lymphocytic leukemia. *Blood* 109.4944–51

55 Esquela-Kerscher, A., F.J. Slack. 2006. Oncomirs – microRNAs with a role in cancer. *Nat Rev Cancer* 6.259–69

56 Fazi, F., A. Rosa, A. Fatica, V. Gelmetti, M.L. De Marchis, C. Nervi, I. Bozzoni. 2005. A minicircuitry comprised of microRNA-223 and transcription factors NFI-A and C/EBPalpha regulates human granulopoiesis. *Cell* 123.819–31

57 Nervi, C., F. Fazi, A. Rosa, A. Fatica, I. Bozzoni. 2007. Emerging role for microRNAs in acute promyelocytic leukemia. *Curr Top Microbiol Immunol* 313.73–84

58 Oscier, D.G., A.C. Gardiner, S.J. Mould, S. Glide, Z.A. Davis, R.E. Ibbotson, M.M. Corcoran, R.M. Chapman, P.W. Thomas, J.A. Copplestone, J.A. Orchard, T.J. Hamblin. 2002. Multivariate analysis of prognostic factors in CLL: clinical stage, IGVH gene mutational status, and loss or mutation of the p53 gene are independent prognostic factors. *Blood* 100.1177–84

59 Eis, P.S., W. Tam, L. Sun, A. Chadburn, Z. Li, M.F. Gomez, E. Lund, J.E. Dahlberg. 2005. Accumulation of miR-155 and BIC RNA in human B cell lymphomas. *Proc Natl Acad Sci USA* 102.3627–32

60 Metzler, M., M. Wilda, K. Busch, S. Viehmann, A. Borkhardt. 2004. High expression of precursor microRNA-155/BIC RNA in children with Burkitt lymphoma. *Genes Chromosomes Cancer* 39.167–9

61 Kluiver, J., S. Poppema, Jong D. de, T. Blokzijl, G. Harms, S. Jacobs, B.J. Kroesen, Berg A. van den. 2005. BIC and miR-155 are highly expressed in Hodgkin, primary mediastinal and diffuse large B cell lymphomas. *J Pathol* 207.243–9

62 Kluiver, J., E. Haralambieva, Jong D. de, T. Blokzijl, S. Jacobs, B.J. Kroesen, S. Poppema, Berg A. van den. 2006. Lack of BIC and microRNA miR-155 expression in primary cases of Burkitt lymphoma. *Genes Chromosomes Cancer* 45.147–53

63 Kluiver, J, Berg A. van den, Jong D. de, T. Blokzijl, G. Harms, E. Bouwman, S. Jacobs,

S. Poppema, B.J. Kroesen. 2007. Regulation of pri-microRNA BIC transcription and processing in Burkitt lymphoma. *Oncogene* 26.3769–76

64 He, L., J.M. Thomson, M.T. Hemann, E. Hernando-Monge, D. Mu, S. Goodson, S. Powers, C. Cordon-Cardo, S.W. Lowe, G.J. Hannon, S.M. Hammond. 2005. A microRNA polycistron as a potential human oncogene. *Nature* 435.828–33

65 Tanzer, A., P.F. Stadler. 2004. Molecular evolution of a microRNA cluster. *J Mol Biol* 339.327–35

66 O'Donnell, K.A., E.A. Wentzel, K.I. Zeller, C.V. Dang, J.T. Mendell. 2005. c-Myc-regulated microRNAs modulate E2F1 expression. *Nature* 435.839–43

67 Sonoki, T., E. Iwanaga, H. Mitsuya, N. Asou. 2005. Insertion of microRNA-125b-1, a human homologue of lin-4, into a rearranged immunoglobulin heavy chain gene locus in a patient with precursor B-cell acute lymphoblastic leukemia. *Leukemia* 19.2009–10

68 Nakae E, Cornell JE, Calin GA, Pollock BH, Croce CM, Yu A Diccianni MB. MicroRNAs in T cell acute lymphoblastic leukemia.[abstract]. 2007. *American Association for Cancer Research Annual Meeting: Proceedings; 2007 Apr 14–18; Los Angeles, CA*. Philadelphia (PA): AACR; 2007. Abstract nr4519.

69 Bullinger L, Sander S, Holzmann K, Russ A, Kestler HA, Pollack JR, Dohner K, Schlenk RF, Dohner H. MiRNA expression profiling based identification of subclasses in adult acute myeloid leukemia.[abstract]. 2007. *American Association for Cancer Research Annual Meeting: Proceedings; 2007 Apr 14–18; Los Angeles, CA*. Philadelphia (PA): AACR; 2007. Abstract nr2967.

70 Ovcharenko D, Kelnar K, Davidson T, Shelton J, Kosel D, Brown D, Illmer T. Role of microRNAs in human acute myeloid leukemia (A ml) samples with different cytogenetic risk profiles.[abstract]. 2007. *American Association for Cancer Research Annual Meeting: Proceedings; 2007 Apr 14–18; Los Angeles, CA*. Philadelphia (PA): AACR; 2007. Abstract nr1779

Keep it simple: microarray cross-platform comparison without statistics

Damir Herman

Myeloma Institute for Research and Therapy, University of Arkansas for Medical Sciences, 4301 West Markham, Slot #776, Little Rock, Arkansas 72205, USA

Abstract

With the proliferation of statistical algorithms developed for analyzing microarray data, high throughput molecular biology has shifted focus to highly numerical data exploration. We are increasingly aware that, although appealing, complicated statistical algorithms cannot remedy all discordance in microarray data, if such exist. In this chapter we explain how significant biological insight can be obtained from a carefully designed microarray experiment where intuition can often replace the need for statistics. We discuss analysis of gene expression data in the context of the FDA spearheaded MicroArray Quality Control project, a comprehensive public effort with its first round of results published in Nature Biotechnology in September 2006. We base our analysis on an understanding of commercial probe-design philosophies and comprehensive probe mapping. With a concrete example, we illustrate the rich biology often overlooked in microarray research and we discuss the merits of cross-platform comparison in clinical setting.

Introduction

According to the Boston Consulting Group (BCG) study [1], to bring a drug from the laboratory to pharmacy shelves in the pre-genomic era required, on average, $880 million and 15 years of research. Currently, costs of carrying out all three phases of a clinical trial commonly well exceed $1 billion [2]. Despite such costly efforts, in the period from 1996–2003 the US Food and Drug Administration (FDA) has seen a steadily declining trend in the number of major drug and biological product submissions [2]. In the same publication BCG conservatively estimates that this staggering amount can be reduced by as much as one-third and the process shortened by two years if genomic technologies are fully utilized.

Microarrays in Inflammation, edited by Andreas Bosio and Bernhard Gerstmayer

Microarray standards

The US Food and Drug Administration (FDA) recently identified "Standards for Microarray and Proteomics-Based Identification of Biomarkers" as its second highest priority for development of new biomarkers and disease models to improve both clinical trials and medical therapy [3]. Despite their great promise, it is believed that microarrays still have not delivered to their full potential [4]. In the absence of microarray standards, the FDA took the initiative to establish a large consortium – the MicroArray Quality Control (MAQC) project – where everything from reagents and arrays to expertise was voluntarily contributed, to investigate the reliability of microarray biotechnology. MAQC was carried out by leading government, academic, and private-industry scientists and is thus far the most ambitious and most comprehensive microarray collaboration. The effort was featured in the September 2006 issue of Nature Biotechnology and this chapter discusses a portion of that extensive work [5, 6, 7].

In the first step, the MAQC consortium [5] measured gene expression in two high-quality RNA samples on the seven commercial microarray platforms available at the time in addition to three alternative quantitative technologies. The two RNA samples were biologically diverse and completely unrelated: Stratagene UHRR (universal human RNA reference, labeled A) extracted from ten different cancer cell lines, and Ambion brain (labeled B), pooled from dozens of Caucasian individuals. Despite criticisms about their choice [8], the consortium selected both RNA samples after carefully evaluating their suitability in a pilot study.

The analyzed RNA samples are still abundantly available, so the MAQC project is one of the rare instances in biotechnology research where data can be easily regenerated and results independently verified. The MAQC project was also unique, in that probe sequences were provided by the majority of manufacturers, which offered a unique view into the probe design process. Contrary to popular belief, the great majority of probes were very carefully chosen and of the highest quality. In section 3, we comment on how availability of probe sequences was used in gene-centered data analysis. With probe sequences in hand, we now understand that careful mapping can resolve typical apparent discordances that have made microarrays infamous.

Despite the explosion of interest in high-throughput gene expression technologies during the past decade, microarray standards have remained elusive, but the MAQC project has resulted in the closest we currently have to a microarray standard. In addition to samples A and B, the researchers also assessed their mixtures in ratios of 3:1 (labeled C) and 1:3 (labeled D). Except for alternative expression technologies such as PCR, every MAQC sample was run in five technical replicates across three independent test sites, generating an unprecedented abundance of identical expression data [9].

The MAQC work was a snapshot in time of a rapidly developing field. Transcriptional databases evolve, as does probe design and technology, so this extensive comparative data will soon be of historical value. One way of continuing to utilize this rich set would be to set it as a benchmark to gauge laboratory and platform performance. This is important because it makes data comparison straightforward for non-obsolete platforms. For example, Affymetrix U133 Plus 2.0 chip, employed by MAQC, is still widely used in basic research geared toward clinical practice, and MAQC Affymetrix data can be used to compare individual laboratory performance. For obsolete platforms that are being phased out and replaced by newly upgraded versions, comparison against the MAQC data is not as trivial, but careful probe mapping usually can be of great help. That is not the case if standard operational procedures (SOPs) such as chemistry and amplification processes drastically change.

Analysis of microarray data

Development of high-throughput microarray platforms profoundly changed molecular biology, where huge data sets have replaced relatively small quantities of data in a relatively short time. During that transitional period, traditional molecular biologists often seemed overwhelmed by the large amounts of generated data. Following the impulse to bring statisticians on board generally led to successful outcomes, but also resulted in a proliferation of data analysis methods. As these data analyses became more complicated and the focus on biology started shifting, it was clear that creative statistical analysis could not remedy all experimental problems.

With that in mind, this chapter focuses on data analysis methods that are extremely simple and intuitive. We show how going back to basics can boost confidence in seemingly hopeless and incoherent data: when we carefully reinterpret data using all information available in the public domain, microarray concordance is evident and it becomes obvious that microarrays indeed measure biological effects rather than noise. Moreover, with extensive supporting evidence from other gene-expression platforms, it is trivial to pinpoint potentially problematic cases that would traditionally be intractable.

Data analysis should be biologically motivated, so we completely utilized all the available information, including public and confidential sources for which we had to sign non-disclosure agreements with platform providers. Data analysis was primarily based on the provided probe sequences, and results were derived from the better understanding of the platform design process that prompted the majority of microarray manufacturers to place their probe sequences into the public domain (if that was not their prior practice) or make them available to customers upon request.

Next, we list major transcriptional databases and give a high level overview of the probe design process. We comment on MAQC's choice of using the transcriptional lowest common denominator database in their cross-platform comparison.

We offer a simple, intuitive and biologically justifiable data analysis algorithm that naturally follows from experimental design. We discuss potential problems of perceived microarray discordance and offer a detailed explanation for embryonic lethal, abnormal vision, *Drosophila*-like 1 (Hu antigen R) or ELAVL1 (GeneID 1994), a gene that according to NCBI Entrez Gene "destabilizes mRNAs and thereby regulates gene expression". This gene is briefly mentioned in the MAQC Nature Biotechnology publication on the comparison between microarrays and alternative quantitative expression technologies such as qRT-PCR [6], but the article does not describe the exciting biology which may be one of the key reasons for microarray infamy.

Finally, in the last section, we conclude by discussing the feasibility and usefulness of cross-platform comparative studies in clinical settings. It is unrealistic to believe that precious biopsy samples can be run in dozens of technical replicates across scores of gene expression technologies. However, knowledge gained from analyzing MAQC data may clearly indicate which haystack to search first when looking for a potentially missing needle.

Some major commercial arrays and their probe design

Since the inception of microarray biotechnology in the mid 1990's [10], microarrays have undergone an evolution of their own. In those early days, a *Saccharomyces cerevisiae* complete genome of "only" 12,068 kb was known [11], and human genome and transcriptional databases were still in their official infancies. In parallel with the human genome project, we witnessed proliferation of sequencing and sequence data collection projects.

In the late 1990s, National Center for Biotechnology Information (NCBI) officially spearheaded a public effort to provide tools for web-based exploration of rapidly growing sequence data beyond "traditional flatfile views" [12]. At that time it was difficult to appreciate the breadth and scope of the probe-design process because probe sequences were determined against rapidly moving targets and were rarely made available to the wider scientific community. For the most part this is still the case today: due to new sequencing projects, expanded knowledge and extensive curation, transcriptional databases constantly evolve. Furthermore, complete probe sequence information is still not customarily offered with generally expensive transcriptional kits, but that situation has recently improved significantly.

Major transcriptional databases

To understand probe-design philosophies for some of the most widely used commercially available microarray platforms, it is worth touching upon a handful of major transcriptional databases. The choice is not meant to be exhaustive but, rather, illustrative of the complexity involved in the research-design-marketing pipeline. Unfortunately, microarray technology has no standardization in any regard, which is the easiest reason to blame it for perceived discordance across different platforms. Once we understand the issues, it will become clear that we must reformulate the terms in which we think about these high-throughput phenomena.

GenBank

GenBank, the first-stop public sequence-deposit database, has followed Moore's Law [13] from its start, doubling every 18 months; in 2006 it received 15 million sequences. The influx of whole-genome shotgun data, along with several versions of the human genome including the first diploid release [14], is primarily responsible for the total collection topping some 145 billion bases. The most comprehensive nucleotide database, GenBank routinely exchanges data with the European Molecular Biology Laboratory (EMBL) [15] and DNA Data Bank of Japan (DDBJ) [16].

The large quantity of deposited data comes with the price of redundancy and lower quality. For example, NCBI human genome reference assembly (version 36.2 as of this writing) is the cleanest publicly available sequence database, yet it is estimated to have an error rate of 10^{-4}. Without going into the details of the genomic pool used in the reference assembly, we expect that roughly every 10,000 nucleotides will have an incorrectly called base.

dbEST

Expressed sequence tags, or ESTs, are several hundred nt long "single-pass" cDNA sequences. The sequencing error rate is estimated at 3% [17], which is more than two orders of magnitude poorer than that of the human genome, and their quality of ESTs may further be diminished by vector contamination. At the time of their deposition into dbEST, an EST division of GenBank, it was often unclear whether the ESTs came from coding or non-coding regions. Moreover, their orientation may pose particular challenges, with far reaching consequences. For example, a number of clinical cancer papers (references deliberately not provided) base their main conclusions on probes designed against ESTs that are incorrectly aligned on the opposite DNA strand. Such probes can, at best, be used to estimate the experimental background; however, if they show strong expression we either get a glimpse into cross-hybridization, a still open field in understanding microarray data, or some new exciting biology.

UniGene

The next level of increasing complexity involves association of ESTs into clusters that correspond to particular transcriptional loci that ideally would represent individual genes. UniGene [18] is a project run by NCBI that gathers all similar ESTs with no attempt to clean them against the genome. Redundancy is still a problem at this level of granularity because, as of August 2007, UniGene build #206 recognizes almost seven-million sequences spread across over more than 100,000 clusters. It is not uncommon for these clusters to get withdrawn as new transcriptional knowledge emerges daily.

GenBank mRNAs

From the vast quantities of deposited sequences, it is reasonable to expect some high-quality ESTs to perfectly align on the genome without any necessity for curation. Such high quality transcripts have been successfully used in studies of alternative splicing, especially for microarray data [19].

RefSeq

In terms of quality, the expertly curated and maintained Reference Sequence database, popularly known as RefSeq [20] is at the top of this short, NCBI biased, list of transcriptional databases. RefSeq, with an error rate of 10^{-3} is a non-redundant, comprehensive, high-quality public database built from all sequences deposited in GenBank. It is important to point out that RefSeq transcripts are not transcribed in cells, they are the result of curation from abundant transcriptional evidence. In that sense, RefSeq transcripts play the same role for GenBank as review articles play for peer-reviewed literature: they are based on a plethora of summarized and concise research information [21].

The high quality comes at a different price – at some 24,000 transcripts, RefSeq is still incomplete. Moreover, with, on average, four transcripts per every three genes, alternative splicing is still an open problem which is effectively addressed by annotating the longest possible transcripts.

RefSeq has several levels of quality and curation but, for the purpose of this discussion, we distinguish only between the lower quality model transcripts (prefixed XM) and higher quality mRNAs (prefixed NM) for which extensive transcriptional evidence and, usually, the coding protein are known.

High-Level Probe Design

With this operational knowledge of major transcriptional databases, we turn to high-level probe design for the commercial microarray platforms used for the MAQC study. Exact algorithmic details are proprietary, but common themes run among all of them. The most notable commonalities are the number of probes (ranges between 20,000 and 50,000), extensive filtering and screening for repeats, penalizing probes that are too far from the 3' end, and comprehensive wet-validation.

ABI, Applied Biosystems one color Human Genome Survey Microarray v 2.0
According to the published white paper [22], for a starting point to develop their whole human genome microarray, ABI used human Celera Transcripts from Celera Genomics Human Genome Database, 98% of RefSeq, cDNA sequences from the Mammalian Gene Collection (MGC) [23], GenBank mRNAs and transcripts experimentally validated in-house. The result was close to 33,000 probes mapping to some 28,000 genes. All probes are 60 nucleotides (nt) long, pre-synthesized and individually placed on the array.

The claim that almost 8000 transcripts and over 25% of the genes on the ABI chip are not in the public domain appears to give this microarray a considerable advantage. For discovery research and cross-platform comparison, however, records not in the public domain are of little practical importance because neither their sequences nor accompanying literature on the context in which those genes are likely to appear are available.

AFX, Affymetrix HG-U133 Plus 2.0 GeneChip
Probe design for the AFX microarray is based on UniGene release 133, as indicated by the "U133" in its name. Affymetrix offers a unique platform because for every targeted sequence, called a sif (named after Sequence Information File in the Affymetrix library), at least one set of 11 or more 25 nt long probe pairs aligned along that target: one set consists of all perfect matches (PM) and another set with deliberately carries a complemented middle base (mismatched probes, MM). With that many probes per target, probeset-level signal expression summary for Affymetrix arrays is both art and science and over 30 normalization methods [24] are currently known. Probes are lithographically grown, which is well understood and commonly used in semi-conductor industry.

AG1, Agilent one-color (Cy3)Whole Human Genome Oligo Microarray, G4112A
Agilent initially offered this microarray in a two-color version, but also introduced a one-color version. The first-tier design databases include RefSeq, Ensembl, University of California Santa Cruz Golden Path and proprietary Incyte [25]. Target sequences are grouped into GeneBins from which consensus regions are generated; 60-nt probes were then designed against these GeneBins. Based on customer request and interest, second-tier input, such as UniGene and other GenBank accessions, are used. Probes are manufactured *in situ* with the Agilent

ink-jet printer technology after extensive *in silico* validation. Wet validation consists of picking probes that exhibit consistent differential expression across ten different samples.

GEH, GE Healthcare CodeLink Human Whole Genome, 300026
CodeLink probes are 30-nt long and were designed against NCBI UniGene #165, RefSeq and dbEST, all released in January 2004 [26]. To avoid potential cross-hybridization, probes are at most 90% similar. They are filtered based on GC content, mononucleotide repeats, melting temperature, stem-loop and secondary structure, and distance from the 3' end. Probes that are more than 2000 nt away from the 3' end are discarded.

ILM, Illumina Human-6 BeadChip, 48K v1.0
Illumina offers 50-nt probes with design tied entirely to RefSeq release from April 2004 [27]. The bulk of these probes are designed against curated NM transcripts, but Illumina also uses model XM transcripts and UniGene #163 from October 2003.

MAQC probe mapping transcriptional database lowest common denominator

As we pointed out, all these transcriptional databases and their versions are moving targets, and from this discussion the best strategy for cross-platform comparison is not clear. For any particular choice, we would introduce a bias in the analysis, as one platform would benefit at the expense of another. In the absence of a standardized probe-design procedure, the MAQC consortium decided to move forward by selecting the RefSeq release from March 2006. From our careful mapping procedure [28], that looked for perfect matches on NM-prefixed transcripts only, we found that roughly 40% of probes on each whole genomic platform routinely covered over 90% of RefSeq, even though arrays were designed 2–3 years prior to our RefSeq freeze. Under the established rules and fixing the transcriptional database release, probe mapping is one of the few exact problems in computational biology that is exactly reproducible.

It was evident that private-industry research and development efforts effectively predated the careful public curation process by several years. We also understand that basic questions addressed by the MAQC consortium, such as microarray reproducibility, had been thoroughly investigated by manufacturers during their chip design process; however, because of commercial interests, none of these results had been publicly scrutinized. MAQC came in to fill this gap in public knowledge.

Who needs statistics? Data analysis 101

What does it mean to have a platform with 25,000 genes and compare it to another platform with 30,000 genes? Does an extra 5000 genes make a significant difference? How do we handle situations in which we have more than one probe per gene? Are the extra probes gene specific (i.e., placed in the common region for all known alternatively spliced variants), or do they target a particular isoform? If they target the same splice variant, could we expect discordance from biological differences of interrogation regions on these transcripts? We now answer all these questions, but go into detail only for the last one.

Is the best platform the one with most genes?

In discovery research no whole genomic microarray platform has a clear advantage because the goal is to unveil a disease-related genetic fingerprint on the global genomic scale. Roughly 60% of probes on the commercial whole genomic microarray do not hit RefSeq. The manufacturers annotate the great majority of these as "exploratory", with hardly any additional transcriptional support, coded protein or published literature. Despite our limited knowledge about the identities of the intended targets, such probes tend to show differential expression to the extent the RefSeq annotated probes do. In other words, there are many unexplained but highly reproducible biological effects beyond our current understanding. That fact should not hinder further research into the applicability of microarray biotechnology in personalized diagnostics or risk assessment, because sets of probes of the same or similar predictive power are known to be non-unique [29, 30]. Therefore, the apparent advantage of extra thousands of profiled genes usually does not lead to any additional insights.

For therapeutics development, questions of fine-mapping details can rarely be answered with manufacturer-provided annotation. Every detail matters when zeroing in on a tractable subset of probes believed responsible for the clinical phenotype under investigation. Finding these details requires careful consideration, as different gene-centric or transcript-centric philosophies of probe design can rarely be inferred from annotations. In addition, data interpretation is inherently time dependent: what may be plausible today may not hold tomorrow due to the emerging transcriptional evidence submitted to GenBank. It may also be possible that gene-specific probes do not corroborate transcript specific results, which is a strong case for cross-hybridization.

More than one probe per gene

From a statistical stand-point, more than one probe per gene usually presents artificial conceptual problems in data analysis. Probes intended for the same targets may not show expression concordance because of differences in probe sequence content. It is very common that a handful of probes out of typically 11 in Affymetrix probesets never get expressed, yet we can have measurable overall levels of transcription.

There are as many examples where we can pinpoint biological reasons for the perceived discordance exactly as there are examples where data interpretation is subject to speculation. For instance, we must consider effects out of our control, such as regulatory elements along the target transcripts, sub-optimal melting temperatures, and secondary structures, as well as carefully attend to factors within our control, such as strict adherence to standard operational procedures. Statistics cannot be the panacea for solving these problems. Consider the extreme example of a population that consists of equal numbers of vegetarians and strict carnivores. On average, the entire population has a balanced diet, but in our new era of personalized medicine, hyperbolically speaking, we should be striving to use microarrays to unambiguously discern diet preferences for every single individual. This task, although non-trivial, is achievable not by discarding "redundant" probes, but by careful consideration of every available piece of data.

Data analysis based on experiment design

As we mentioned in the Introduction, the whole-genomic platforms used by the MAQC consortium assessed two high quality RNA samples [5] and their mixtures:

A · · · Stratagene Universal Human RNA Reference sample, a mixture of RNA from 10 human cancer cell lines;,
B · · · Ambion Brain, a pool from several dozen primarily Caucasian individuals;,
C · · · $C = 0.75\ A + 0.25\ B$; and
D · · · $D = 0.25\ A + 0.75\ B$.

It is obvious that sample C is more similar to A, whereas sample D should more closely resemble B. All samples were analyzed in five technical replicates across three different test sites. Alternative, non-microarray expression technologies were run at least in triplicate on the manufacturer's test site only.

Let us consider a situation in which a particular probe or probe set exhibits stronger expression in sample A then in B. What do we expect from samples C and D? Because of possible saturation, we should not expect a linear response, but we should undoubtedly see a monotonous response $A > C > D > B$ [7]. In the opposite case when $A < B$, it must hold that $A < C < D < B$. If we plot signal intensity as a function of sample, where we order biological samples as A, C, D and B, regardless of the technical replicates' order within the sample, we expect a monotonous line.

When we put all test sites on the same plot, we expect to see a qualitatively identical saw-tooth pattern on all three test sites; any other outcome reveals a problem.

Particular example: gene ELAV1

To illustrate this point, we consider all MAQC probes mapped onto embryonic lethal, abnormal vision, Drosophila-like 1 (Hu antigen R) gene or ELAVL1, GeneID 1994.

Problem

Preliminary data analysis showed a typical discordance present in micrarray data analysis: different probes mapped to this gene on the same microarray platforms exhibited qualitatively different behavior. This gene was presented as one of the main expression data conundrums during the MAQC conference in Palo Alto in December of 2005.

Mapping at the probe annotation level

In addition to the commercial platforms described in the previous section ("Some major commercial arrays and their probe design"), we also take into consideration the available MAQC data from the whole-genomic National Cancer Institute (NCI) custom-printed Operon arrays [5] and three alternative expression technologies: Panomics QuantiGene, Applied Biosystems TaqMan, and Gene Express (Sta)RT-PCR. QuantiGene is the only technology capable of directly measuring RNA without any amplification, and the other two are PCR based [6].

We mapped the representative probes, probe sets and assays for the discussed MAQC platforms on gene ELAVL1 over its RefSeq representative transcript NM 001419.2. The target RefSeq transcript is 6058 bp long, has 6 exons, coding sequence stretches from 168 to 1148 and the 3' UTR is 4909 bp long.

Table 1 shows the mapped relation to the extent manufacturers would provide in their annotation files [31]. From this table it is not clear at all where the matches occur, and why would three out of five commercial manufacturers (Affymetrix, Agilent and GE HealthCare) design more than one probe for this gene?

Fine resolution mapping

To simplify the subsequent discussion, we go into fine resolution mapping, presented in Table 2. Table 2 provides interrogation start and end positions on the target RefSeq NM 001419.2 transcript, distance from the 3' end, and the relevant targeted length. For Affymetrix, we list the start position of the 5' most probe and end posi-

Table 1 - Mapping results for Gene ELAVL1, GeneID 1994, RefSeq transcript NM 001419.2,gi 38201713. ABI stands for Applied Biosystems, AFX is Affymetrix, AG1 is Agilent 1-dye, GEH is GE HealthCare, ILM is Illumina, TAQ is ABI TaqMan, QGN is QuantiGene and GEX is Gene Express.

ABI	214864
AFX	201726_at; 201727_s_at; 227746_at; 244660_at
AG1	A_23_P208477; A_23_P388681; A_32_P200165
GEH	GE59120; GE787419
ILM	GI_38201713-S
NCI	H200002121
TAQ	Hs00171309_m1
QGN	PA-11191
GEX	ELAVL1

tion of the 3' most probe in the corresponding probe set. ABI provided the 25-nt probes in the amplified regions but was unwilling to provide locations of TaqMan primers, so we merely corroborated TaqMan annotation; therefore, we could not specify the exact length of the intended target.

From Table 2 it is evident that QuantiGene assay PA-11191, Affymetrix probeset 201727_s_at, Gene Express ELAVL1 and very likely ABI TaqMan Hs00171309_m1 PCR assays interrogate the coding region, whereas all other microarray probes hit in the 3' UTR. Because of the 3' bias inherent in the Ebberwine amplification during cRNA target preparation, we have no reasons to *a priori* expect any signal from probes more than 2000 bp from the 3' end. However, except for Illumina, all MAQC commercial microarray manufacturers, and the custom printed NCI oligo arrays target the region more than 4 kb upstream from the 3' end, with Affymetrix and Agilent doing it twice. A gap of 3 kb between Agilent probe A 23 P208477 and Affymetrix probeset 227746 at is apparent.

Most 5' data

Figure 1 shows the expression data for the first ten probes, sorted according to their start position relative to the 5' end, as presented in Table 2. The first five probes mapped to the coding region and the next five target up to roughly 900 bp into the 3' UTR. For clarity, GEX, TAQ, GEH and NCI results were offset by -1, 3.25, 5 and -2 on the $\log_2$ scale, respectively.

From the data, we can conclude several things. First, all platforms, except Gene Express (magenta line) and GE HealthCare (orange solid lane) show ascending monotonous behavior. Gene Express assay ELAVL1 clearly fails: A > B, so C should

Table 2 - Mapping details with hits to gene ELAVL1, GeneID 1994, RefSeq transcript NM 001419.2, gi 38201713 sorted according to their start position from the 5' UTR. Target RefSeq transcript is 6058 bp long, has 6 exons, coding sequence starts at position 168, ends at 1148, and the 3' UTR is 4909 bp long. Hit Start is the 5' most position and Hit End is the 3' most position perfectly matched on the transcript. Hit length refers to the stretch of mRNA targeted by the assays, probes or probe sets.

	Probe/Assay	Hit Start	Hit End	Distance to 3'	Hit Length
QGN	PA-11191	236	662	5823	497
AFX	201727_s_at	733	1240	5326	508
GEX	ELAVL1	801	1297	5258	497
TAQ	Hs00171309_m1	<810	>834	<5249	>25?
GEH	GE59120	1034	1063	5025	30
AG1	A_23_P388681	1177	1236	4882	60
ABI	214864	1210	1269	4849	60
NCI	H200002121	1433	1501	4626	69
AFX	201726_at	1858	2255	4201	398
AG1	A_23_P208477	2010	2069	4049	60
AFX	227746_at	4848	5280	1211	433
AG1	A_32_P200165	5246	5305	813	60
GEH	GE787419	5445	5474	614	30
AFX	244660_at	5736	6003	323	268
ILM	GI_38201713-S	5893	5942	166	50

be bigger than D, but we measure $C < D$ which is quite the opposite. GE HealthCare probe GE59120 does not titrate at all. Because we chose not to calculate error bars, we immediately spot a problem with Agilent probe A_23_P388681 (sold red line) on site 2, sample D, replicate 2. It is also obvious that the 3' more Agilent probe, A_23_P208477 (dashed red line), does show a monotonous but not a linear dose response – it is saturated for 100% and 75% of Stratagene UHRR (samples A and C) and then precipitates at D and especially B. Affymetrix probe sets 201727_s_at (solid green line) and 201726_at (dashed green line) are perfectly correlated. The later, more 3' Affymetrix probe set 201726_at is more strongly expressed, which is a classical example of the 3' bias, a well known shortcoming of non-random primed microarray biotechnology. The 3' bias is further corroborated by the Agilent probes. A problem with NCI probe H200002121 on site 2 (lot of missing data) and site 3 (strong variability for sample D) is also obvious. The remaining alternative technologies, QuantiGene and TaqMan, march in lockstep, as shown by their assay data QGN PA-11191 and TAQ Hs00171309_m1.

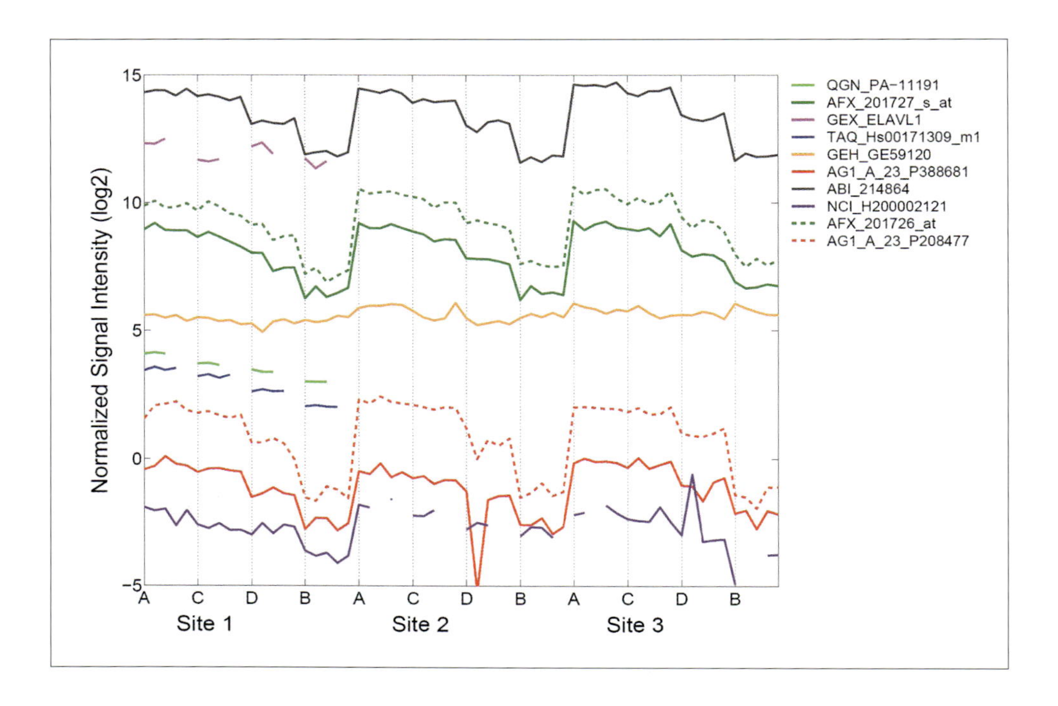

Figure 1.
Plot of the first ten 5' most mapped probes from Table 2.
Vertical axis is log_2 normalized signal. Along the horizontal axis, we split the data according to the three test sites, which we further refine according to the relative concentration of A sample, with five technical replicates each. All probes that follow dose response indicate that sample A is more strongly expressed than sample B. Note that lines are connected as we move from one test site to another, which is a plotting artifact. Colors: QGN, solid light green AFX, solid and dashed dark green; GEX, magenta; GEH, solid orange; AG1, solid and dashed red; ABI, black; NCI, purple.

Qualitative Difference – Most 3' Data

Further along the RefSeq transcript NM 001419.2 towards the 3' end, the last 5 mapped probes mapped within 1500 nt from the 3' end (Tab. 2) qualitatively behave in a completely opposite way: now $A < B$, indicating stronger expression in the Universal Reference RNA than in brain. We cannot ignore this behavior because all platforms consistently show the same effect. How do we resolve this apparently discordant behavior?

From our previous discussion on probe design and the 3' bias in the section "Some major commercial arrays and their probe design", we recall that GEH

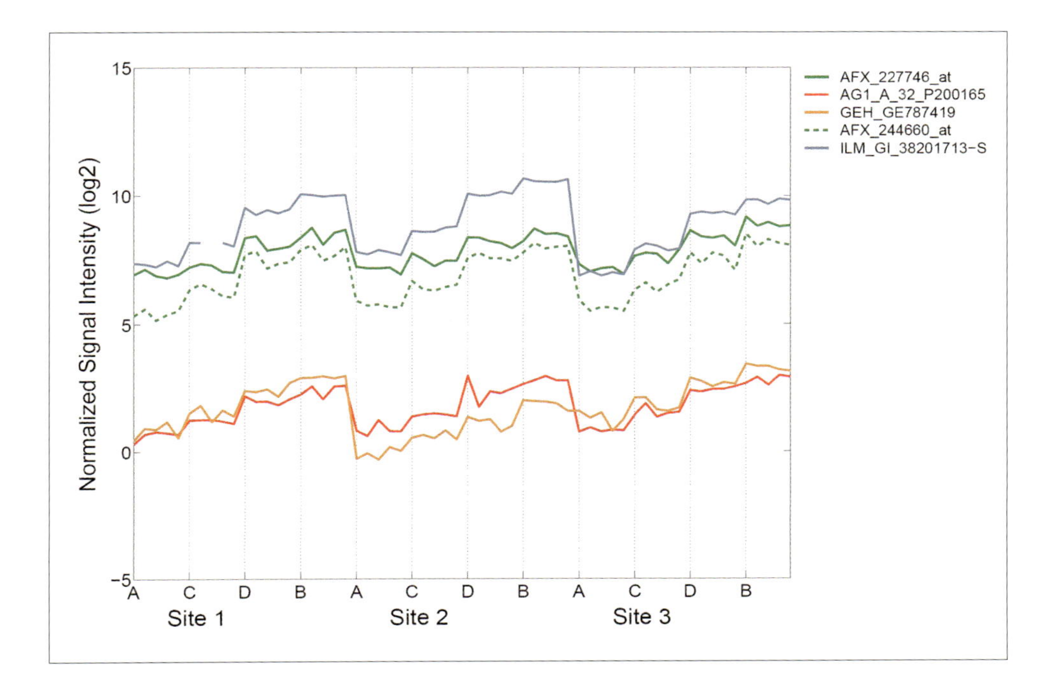

Figure 2.
Probes mapped on gene ELAVL1, GeneID 1994, to 1500 and less nt from the 3' UTR on NM_001419.2. In comparison with Figure 1, we see completely opposite qualitative behavior, where now B is more strongly expressed than A. All probes are perfectly correlated and show linear dose response. Colors: AFX, solid and dashed green; AG1, red; GEH, orange; ILM, gray. Note the outlier for Agilent probe A_32_P200165 on site 2, sample D, replicate 1.

assigns no score to probes that map further than 2000 nt from the end of the 3' UTR. It may be surprising that we get signal way deep in the 5' region at over 5000 nt from the 3' end, but we see that probe design at these regions is not a mistake but a deliberate act practiced by all commercial whole genomic platform providers, with the exception of Illumina.

Resolution: Biology or error?

Careful reading of the NCBI curation record for NM 001419.2 [32] reveals that there are two polyadenylation [33] sites: one is the usual suspect at the very end of the 3' UTR, but there is also another one at position 2358! That piece of NCBI annotation completely explains why we see any differential expression so far 5' and also leads us to the conclusion that brain expresses the same ELAVL1 isoform with much longer UTR. Similar analysis for other genes in the MAQC study indicates that holds true, in general, for brain samples.

It should be obvious from this example that no statistics would be able to explain this purely biological effect. A coarse statistical analysis would merely discard this type of discordance, or even worse, take the average over all expressed values. We see that data analysis based on probe mapping and experimental design can offer insights that standard figures of merit such as fold changes and p-values cannot.

The take home message of this detailed example is that there are alternatives to purely statistical data analysis. Because of the pace at which sequence data is collected and curated, it is becoming increasingly easier to understand and interpret results of microarray experiments. Consequently, we are more likely to see many of the problems associated with microarray technology resolved in the near future.

Cautiously forward from bench to bedside

Without a golden standard, what do we do?

Due to the absence of standardization in microarray-based research, we cannot assert with absolute certainty that what our microarray study measures is the real biological effect. From the current state of expression technology, we cannot even assess how much of our data is skewed by the technologically caused artifacts. The best one can do is to amass enough corroborating evidence to support the experimental hypothesis. A common example is the PCR "validation" on a handful of investigated genes. From Figure 1, however, it is evident that PCR is not a golden standard and, according to the presented data analysis, it may fail (Gene Express assay, pink line). This is an undesirable but unavoidable consequence of the expression technology. Such a failure rate, as present in the MAQC data at a few percent, is comparable with microarray performance. Even if both technologies perform to the extent that they increase our confidence in the results, that does not mean we can measure gene expression more accurately than a person with two clocks can measure time when neither is atomic.

Microarray hybridization is a highly non-trivial process that requires considerable effort and a high level of wet bench proficiency. The rather involved lab procedures require a significant number of steps that we believe multiplicatively contribute to the overall noise in the results. Based on our MAQC work on artificially discrepant samples, Ambion brain and Stratagene universal RNA, we now know that microarrays can be highly reproducible. The academic insight gained from the MAQC project and its plethora of technical replicates is invaluable for any clinical setting, since similar design is very remotely applicable to real clinical situations.

Technical replicates in clinical research

It is highly impractical to run enough technical replicates to satisfy every statistician. RNA extracted from clinical patients is too scarce and to valuable to be used in the way that would please basic scientists with little experience in clinical research. As a concrete example, to provide tissue samples for research, our patients with myeloma voluntarily subject themselves to a very painful, uncomfortable bone marrow extraction procedure. Depending on the patient's disease progression stage, the extracted bone marrow may provide only a limited number of cells that yield RNA quantities sufficient for a single microarray experiment, let alone multiple experiments.

On usefulness of probe mapping

When establishing a diagnostic personalized fingerprint from genetic profiling for each individual patient, it is sufficient to stop at probe-level data. There is no need to take a leap forward and plunge into sketchy pathway analysis. Usually, manufacturer-provided annotation data is correct to a great extent, but for biological understanding of clinical results it is highly insufficient to follow the commonly practiced "1 probe per gene" maxim. That approach does not work across the board, as different platform providers have different probe-design strategies.

In RefSeq, on average, we have three genes for every four transcripts, which does not appear problematic from a statistical point of view. Our analogy of RefSeq transcripts with review articles reminds us that, just as a review article cannot cover all facets of a research problem, RefSeq does not offer resolution of all mapping issues. Despite its high quality, RefSeq is inherently biased because it tends to ignore intronless transcripts and to provide extensive annotation information for long transcripts that encode proteins with more than 100 amino acids. Most RefSeq revisions consist of extensions on both 5' and 3' ends. Considering the 3' bias limitation of microarray biotechnology, although all acquired data on the probe level is etched in stone, interpretation of gene- and transcript-centered data is time dependent. That state of affairs is not acceptable in the clinical setting.

In clinical practice, multiple platform comparison beyond the realm of basic science is carried out very rarely, if at all. In therapeutic development, it is highly desirable to have complete knowledge of probe sequences, because probe mapping is an important component of data interpretation. Some reports in the literature went as far as mapping probes on the same exon [35], but it is not necessary to go to this level of detail, as we showed in the previous section. We should, however, be aware of long UTRs and alternative polyadenylation sites in target transcripts. We should also consider probe-specific thermodynamic differences (e.g., such as melting temperature directly related to probe GC content) that lead to different hybridization affinities. These effects are highly reproducible, and the MAQC data afford some basic understanding of cross-hybridization. For example, we can safely discard several nucleotides on both ends of probes and still expect a strong signal, as long as the middle 90% or more nucleotides on a probe perfectly match the present target.

Keep it simple!

Although sophisticated statistical and artificial intelligence algorithms have been developed for microarray data analysis, the best advice that should be routinely practiced is to keep the diagnostic part as simple as possible [36]. For a typical study with samples (in the hundreds) greatly outnumbered by the large number of probes (tens of thousands), it is easy to over-fit the data. In addition, the more complicated the model, the harder it is to develop intuition about it.

In the hunt for therapeutic targets, we strongly believe one must go beyond microarrays and corroborate findings with alternative molecular biology techniques, such as florescence *in situ* hybridization to detect cytogenetic abnormalities, array comparative genomic hybridization to zoom in on copy-number variations and miRNA to shed light on expression regulation, to name a few. Moreover, understanding the correlation between gene and protein expression, especially in cancer progression or remission, may hold the key to pushing forward the frontier of personalized medicine.

This entire process is complicated enough, and the key thing is to keep it simple.

Acknowledgments

This work would have been impossible without the support of my mentor and friend, David Landsman, at NCBI. I am infinitely grateful to have had the great privilege of close collaboration with 135 scientists on the MAQC consortium, from whom I learned a great deal about intricacies of gene expression biotechnologies. I would especially like to thank Leming Shi, whose leadership was the key to the success of the project, Jean and Danielle Thierry-Mieg for their infinite patience and guidance, and Richard Shippy for our interaction on the titration project.

I am grateful to Stratagene and Ambion for providing the high quality RNA samples. I am indebted to ABI for making available some thousand TaqMan assays. I would also like to thank the management and R & D experts from ABI, Affymetrix, Agilent, Ge HealthCare, Illumina, Operon and Eppendorf for providing their microarray platforms, including probe sequences and design details. I very much appreciate the guidance I received from GeneExpress and Panomics on my conquest of PCR and non-PCR based genetic profiling techniques, by providing their platforms, assay sequences and expertise.

Finally, I would like to thank Josh Epstein and Peggy Brenner for critical reading of this chapter and their comments.

References

1 Tollman P, Guy P, Altshuler J, Flanagan A, Steiner M (2001) How genomics and genetics are transforming the biopharmaceutical industry, Boston Consulting Group publication, http://www.bcg.com/impact_expertise/publications/files/eng_genomicsgenetics_rep_11_01.pdf
2 Innovation or Stagnation? Challenge and Opportunity on the Critical Path to New Medical Products, http://www.fda.gov/oc/initiatives/criticalpath/whitepaper.pdf
3 Critical Path Opportunities List, FDA's Critical Path Initiative report, http://www.fda.gov/oc/initiatives/criticalpath/reports/opp_report.pdf
4 Marshall E (2004) Getting the noise out of gene arrays. *Science* 306(5696): 630–1
5 MAQC Consortium; Shi L, Reid LH, Jones WD, Shippy R, Warrington JA, Baker SC, Collins PJ, de Longueville F, Kawasaki ES, Lee KY et al (2006) The MicroArray Quality Control (MAQC) project shows inter- and intraplatform reproducibility of gene expression measurements, *Nat Biotechnol* 24(9): 1151–61
6 Canales RD, Luo Y,Willey JC, Austermiller B, Barbacioru CC, Boysen C, Hunkapiller K, Jensen RV, Knight CR, Lee KY et al (2006) Evaluation of DNA microarray results with quantitative gene expression platforms. *Nat Biotechnol* 24(9): 1115–22
7 Shippy R, Fulmer-Smentek S, Jensen RV, Jones WD, Wolber PK, Johnson CD, Pine PS, Boysen C, Guo X, Chudin E et al (2006) Using RNA sample titrations to assess microarray platform performance and normalization techniques. *Nat Biotechnol* 24(9): 1123–31
8 Liang P (2007) MAQC papers over the cracks. *Nat Biotechnol* 25: 27–28
9 GEO accession GSE5350, http://www.ncbi.nlm.nih.gov/geo/query/acc.cgi?acc=GSE5350
10 Schena M, Shalon D, Davis RW, Brown PO (1995) Quantitative monitoring of gene expression patterns with a complementary DNA microarray. *Science* 270(5235): 467–70
11 Goffeau A, Barrell BG, Bussey H, Davis RW, Dujon B, Feldmann H, Galibert F, Hoheisel JD, Jacq C, Johnston M et al (1996) Life with 6000 genes. *Science* 274(5287): 546, 563–7
12 Tatusova TA, Karsch-Mizrachi I, Ostell JA (1999) Complete genomes in WWW Entrez: data representation and analysis. *Bioinformatics* 15(7–8): 536–43
13 Benson DA, Karsch-Mizrachi I, Lipman DJ, Ostell J, Wheeler DL (2007) GenBank, *Nucleic Acids Res* 35 (Database issue): D21–5
14 Levy S, Sutton G, ng PC, Feuk L, Halpern AL, Walenz BP, Axelrod N, Huang J, Kirkness EF, Denisov G et al (2007) The diploid genome sequence of an individual human. *PLoS Biol* 5(10): e254
15 http://www.embl.org, European Molecular Biology Laboratory web site

16 http://www.ddbj.nig.ac.jp, DNA Data Bank of Japan web site
17 Boguski MS, Lowe TM, Tolstoshev CM (1993) dbEST-database for "expressed sequence tags". *Nat Genet* 4(4):332–3
18 http://www.ncbi.nlm.nih.gov/UniGene, National Center for Biotechnology Information UniGene project web site
19 Kahn AB, Ryan MC, Liu H, Zeeberg BR, Jamison DC,Weinstein JN (2007) SpliceMiner: a high-throughput database implementation of the NCBI Evidence Viewer for microarray splice variant analysis. *BMC Bioinformatics* 8:75
20 http://www.ncbi.nlm.nih.gov/RefSeq, National Center for Biotechnology Information Reference Sequence (RefSeq) collection project web site
21 The NCBI Handbook, Ch. 18, http://www.ncbi.nlm.nih.gov/books/bv.fcgi?rid=handbook
22 White Paper: The Design and Annotation of the Applied Biosystems Human Genome Survey Microarray, http://docs.appliedbiosystems.com/pebiodocs/00113304.pdf (requires registration)
23 http://mgc.nci.nih.gov, Mammalian gene collection, full-length open reading frame clones for human, mouse and rat genes
24 http://affycomp.biostat.jhsph.edu/AFFY2/comp_form.ht ml, Rafael A. Irizarry web site with the list of varieties of Affymetrix gene chip normalization, along with examples
25 Kronick MN (2004) Creation of the whole human genome microarray. *Expert Rev Proteomics* 1(1):19–28
26 CodeLink Probe Design, correspondence with GE Healthcare
27 Illumina Human Whole Genome Gene Expression Probe Set, correspondence with Illumina
28 Herman D (2006) BLAST Analysis for the MAQC Project. Reference [5], Supplementary Methods
29 Ein-Dor L, Kela I, Getz G, Givol D, Domany E (2005) Outcome signature genes in breast cancer: is there a unique set? *Bioinformatics* 21(2): 171–8
30 Fan C, Oh DS, Wessels L, Weigelt B, Nuyten DS, Nobel AB, van't Veer LJ, Perou CM (2006) Concordance among gene-expression-based predictors for breast cancer. *N Engl J Med* 355(6):560–9
31 Reference [5], Supplementary Table 2
32 http://www.ncbi.nlm.nih.gov/entrez/viewer.fcgi?val=NM 001419.2
33 http://en.wikipedia.org/wiki/Polyadenylation
34 Jean and Danielle Thierry-Mieg, private discussion
35 Kuo WP, Liu F, Trimarchi J, Punzo C, Lombardi M, Sarang J,Whipple ME, Maysuria M, Serikawa K, Lee SY et al (2006) A sequence-oriented comparison of gene expression measurements across different hybridization-based technologies. *Nat Biotechnol* 24(7):832–40
36 Shaughnessy JD Jr, Zhan F, Burington BE, Huang Y, Colla S, Hanamura I, Stewart JP, Kordsmeier B, Randolph C, Williams DR et al (2007) A validated gene expression model of high-risk multiple myeloma is defined by deregulated expression of genes mapping to chromosome 1. *Blood* 109(6):2276–84

A regulatory perspective

Microarrays in drug development: regulatory perspective

Roland Frötschl and Peter Kasper

Federal Institute for Drugs and Medical Devices (BfArM), Kurt-Georg-Kiesinger-Allee 3
53175 Bonn, Germany

Abstract

This chapter describes the current status and impact of data from microarray applications on drug development and approval from a European regulator's point of view.

Microarray technology is regarded as a new, valuable and increasingly important tool in drug development and risk evaluation. The technique may enable faster development of new and safer drugs, even for diseases as yet incurable. Scientists in industry, academia, and regulatory bodies are currently evaluating and discussing what prerequisites are needed to implement microarray data from pharmacogenetic, pharmcogenomic, and toxicogenomic experiments in the regulatory process and for use in risk/benefit evaluations.

The areas where microarray applications are considered meaningful are identified and the activities and expectations of regulatory bodies discussed. The proposed measures of quality control and assurance are presented as outlined in the draft guidance paper by FDA and a draft reflection paper by EMEA. The current discussion on how to define and qualify biomarkers for prediction of pharmacologic/toxicologic effects, and their implementation in the regulatory framework and decision making is described.

In an overview, regulatory experience with microarray data submissions so far is presented. Our general expectation is that requests for pharmacogenetic briefing meetings and voluntary genomics data submission will increase substantially in the near future, also leading the way to increasing data submissions with marketing applications.

Introduction

"Omics"-technologies started in the '90s of the last century with the development of microarrays and their use in various research applications. There were high expectations for the probable impact of using microarrays e.g. on speeding up development of new drugs for, as yet, incurable diseases, providing mechanistic insights into disease mechanisms, and prediction of adverse effects [1–3]. Some of the initial enthusiasm and over-extended promises have been replaced now by more realistic expectations of what whole-genomic profiling *via* microarrays might contribute to the challenges in the optimisation of drug development and drug safety [4]. How-

Microarrays in Inflammation, edited by Andreas Bosio and Bernhard Gerstmayer

ever, their principal advantage of simultaneous measurement of thousands of genes or expression changes, and thereby delivering data on changes in regulation of functional networks and the whole genome of cells upon a certain challenge, is still regarded as major progress in experimental research capabilities with huge potential for drug development as well as drug therapy and safety [5, 6].

In drug development, microarrays were first implemented in strategies to drug target finding and lead structure identification, and the screening of volunteers in clinical trials for their genetic predisposition. More and more effort has also been put into using microarray technology to find mechanistic explanations of adverse toxic events in animal species and approaches to predicting adverse events from expression profiling data to enhance drug safety. The power of microarray technologies to provide better insight into the molecular mechanisms of a drug's action by identifying transcriptional biomarkers of toxicological and pharmacological effects, and the impact of such data in the regulatory review process of new drugs, is currently a highly discussed topic at the regulatory bodies of Japan, the USA and Europe.

Activities of regulatory bodies

There is an increasing awareness and interest on the part of regulatory agencies world-wide in innovative strategies of integrating transcriptional profiling into the drug development process. Several important regulatory documents on "Genomic-Data" have been published in recent years by regulatory bodies. In 2002 L.J. Lesko and J. Woodcock of the Food and Drug Administration (FDA) of the United States published their paper "Pharmacogenomic-guided drug development: regulatory perspective" [7] where it is emphasized that the FDA was aware of the explosion of Pharmacogenetics and Pharmacogenomics data and the need to discuss their impact on drug approvals with all stakeholders. In its white paper "The Critical Path to New Medical Products" [8] the FDA recognized pharmacogenomics as crucial to advancing medical drug development and personalized medicine. FDA implemented its "Guidance for Industry: Pharmacogenomics Data Submission" [9] in regulatory routine in 2005 and opened the Voluntary Genomic Data Submission (VGDS) path. The Interdisciplinary Pharmacogenomic Review Group (IPRG) has been newly created to review VGDSs. Additionally, guidance on tests for heritable markers (Pharmacogenetic Tests and Genetic Tests for Heritable Markers) was recently issued [10].

In Europe, in 2004 the Committee for Medicinal Products for Human Use (CHMP) of the European Medicines Agency (EMEA) implemented an expert group with the status of a working party, the CHMP Pharmacogenetics Working Party (PgWP). Similar to the VGDS path of the FDA, the EMEA "Guideline on Pharmacogenetics Briefing Meetings" [11] encourages applicants to share information on and experience with genomic-related issues between industry and regulatory agencies. In so-called briefing meetings held at the EMEA, applicants can discuss with PgWP the technical, scientific and regulatory issues that arise from the inclusion of genomic data in development strategy and their potential implications in the regulatory process. Discussions in these forums are not intended to impact ongoing development programs nor do they replace existing expert advice procedures. More recently the European Commission and the FDA agreed to a procedure for joint FDA – EMEA briefing meetings with sponsors following voluntary submission of genomic data [12], which allows applicants from the pharmaceutical industry to obtain the benefit of the expertise, perspectives and ideas available at both agencies for innovative approaches to regulation. FDA and EMEA have held three joint VGDS briefing meetings so far (July 2007) including one that specifically addresses the qualification of novel safety biomarkers. These activities offer important opportunities for companies and regulators to discuss potential applications of genomic approaches and will certainly help build consensus around regulatory policies, quality standards and testing guidances. Different regions (EU, USA, Japan) have published pharmacogenomic and pharmacogenetic specific guidances, or concept papers, and are in the process of developing others. However, the lack of consistently applied definitions to commonly used terminology raises the potential for conflicting use of terms in regulatory documentation or inconsistent interpretation by regulatory authorities. Therefore, the International Conference on Harmonization of Technical Requirements for the Registration of Pharmaceuticals for Human Use (ICH) has recently prepared a guideline which contains definitions of key terms in the disciplines of pharmacogenomics and pharmacogenetics, namely genomic biomarkers, pharmacogenomics, pharmacogenetics and genomic data and sample coding categories [13].

Applications for microarrays

Adverse drug reactions in the clinical development of drugs or lack of efficacy are major reasons for attrition of drugs during clinical development [14]. Early recognition or valid prediction of adverse effects or lack of efficacy would save time and money for companies and enhance safety for volunteers. Pharmacogenetic data from large scale screening for genetic variation up front to clinical trials and/or pharmacogenomic data of gene expression changes in preclinical and clinical development might be powerful tools to predict treatment failures or adverse toxic effects. Predictive validated markers for pharmacogenetic applications are already available [15], whereas predictive models for pharmacogenomics data must still be established [16]. Currently more realistic are mechanistic studies to clarify the mode of action of a toxic effect seen in classical standard tests. In such cases toxicogenomics approaches might be of high additional value to clarify species specificity or dose dependence by identifying the underlying gene groups and networks responsible for the toxic effects and the "no observed effect level" for expression alteration of these genes.

Pharmacogenetic applications

One of the first applications of microarrays in clinical medicine was the determination of HIV-resistance status by simultaneously measuring the single nucleotide polymorphisms (SNPs) in the different genes of viral target enzymes for drugs used in HAART (highly active antiretroviral therapy) [17]. This application rapidly demonstrated the huge value of this highly-parallel screening method in fields where genetic variability is great and of major importance for therapeutic success [18].

Genetic predisposition, e.g. for the activity of specific drug metabolising enzymes such as cytochrome P450 2D6, often results in differences within populations in their rate of response to certain drugs. In drug development it is therefore extremely important to determine the status of genetic variability, especially in genes important for drug activity, metabolism, and transport, which might then be crucial for the activity of a new drug and the responder rate in the relevant patient and/or volunteer groups [18].

Since the announcement of the complete nucleotide sequence of the human genome, the source for SNPs has enormously increased and whole genome SNP scans are already being done and integrated into drug discovery and drug development [19]. For the adequate use of such large scale data and their predictive value, valid biomarkers or surrogate markers or marker profiles for specific phenotypes are necessary [15].

Pharmacogenomic applications

Expression profiling in preclinical stages of drug development is already used by most major pharmaceutical companies to address various issues from target finding, to lead structure identification, and toxicology. The major impact in toxicology is seen in improving the prediction of toxicological profiles of substances and the identification of underlying mechanisms of action responsible for toxicity of drug candidates [20].

For predictive toxicology a reasonable amount of high quality data with reference substances is needed to develop predictive models and to make predictions for new substances. To ensure high quality data collection for reference substances, common standards for data quality and format must be established and accessibility to those databases must be guaranteed.

Expression profiling in clinical practice has been used extensively in diagnostic research of various diseases e.g. breast cancer, ovarian cancer, or multiple sclerosis, aiming to find expression profiles suitable as markers for disease prediction, prognosis, and response to treatment to improve individualized treatment and therapy success [21–23]. This raises hopes for future developments in microarray technology capable of improving drug development by predicting and monitoring the efficacy of drugs and serving as a diagnostic tool for efficient individualized therapy [24]. Especially in relapsing-remitting auto immune diseases, prediction of therapies effective in slowing down disease progression might be extremely beneficial for adjuvant therapeutic approaches. Development of reliable, predictive biomarkers would also have the potential to significantly shorten clinical development of drugs, especially in that therapeutic area.

Regulatory quality standards for microarray data

To become an established part of a regulatory safety testing procedure, test methods and reliability of measured endpoints must be proven in terms of value in revealing a specific biohazard or predicting a certain risk for human use. Quality aspects must be addressed with regard to e.g. reliability, documentation of all relevant experimental steps, number of technical and biological repeats needed, and proof of sensitivity and comparability of the technological platform used at test sites. The definition of the Minimal Information About Microarray Experiments (MIAME) standard was the first step in that direction [25]. With the Microarray Quality Control (MAQC) project finished last year and spearheaded by the FDA, it has been demonstrated that different microarray platforms show a high level of reproducibility within and between platforms, at least when tested with exactly the same set of probes of RNA [26, see also the chapter by D. Herman in this book]. These results have laid a fundamental basis for proceeding further with validation efforts on microarray applications, although critical voices are still heard warning that, despite all progress, the fundamental understanding of the pillars of this technology remain largely unexplored [27].

Quality issues addressed by the MAQC consortium have been transformed into a draft guidance paper [28]. Within this guidance document recommendations are given on methodological issues when generating gene expression data with microarrays and what data should be submitted with those studies. These recommendations cover RNA isolation, handling and characterization with details on:

i) what must be considered before isolation of RNA, such as the use of RNAse-free reagents, RNA-stabilizers, batch sizes, storage conditions and genomic DNA contaminations; .
ii) RNA-isolation from different sources, such as tissues or cells, whole blood or peripheral blood mononuclear cells (PBMCs);
iii) RNA quality control measures, such as UV-absorbance ratio at 260 nm and 280 nm, agarose electropheresis with 28S and 18S rRNA integrity, Agilent Bio-

analyzer generated 28S to 18S ratio, and recommended values of rRNA peak integration, and measurement of RNA integrity number.

The control of labelling reactions is addressed with details on what must be considered before and during labelling, and what data should be submitted to demonstrate adequate labelling performance. Implementation of proficiency testing measurement to avoid procedural failures within the testing facility is highly recommended and suggestions on how this might be achieved are presented e.g. by using universal sets of reference RNAs. Hybridization control is recognized in the concept paper as a critical issue. The current lack of widely accepted QA/QC control metrics is acknowledged, and currently the use of pairs of reference control RNAs is suggested as a possibility. It is recommended that all pertinent information on reproducibility and accuracy of microarray experiments be included in the submission. Detailed advice for microarray scanner settings and the required information that should be submitted are given.

Two additional chapters summarize the critical issues of data analysis leading to the creation of differentially expressed gene lists and the biological interpretation of these lists. It is emphasized that, for the factors listed as having confounding effects on the generation of differentially expressed gene lists, no current consensus criteria exists at this time. Therefore, a clear description of all steps leading to the differentially expressed gene list should be submitted. In the following step of biological interpretation of the data, the critical questions to answer should be addressed as thoroughly as possible with the available analytical tools. These questions include e.g. which and how many pathways are affected, are these pathways tissue specific, and are they relevant for the (proposed) mechanism of action. Similar biological interpretation with different databases may facilitate the building of consensus interpretations. However, each database should be critically reviewed for its quality and relevance, as it is recognised that substantial differences between databases exist. Therefore, it is recommended that all relevant information on the databases used be submitted.

The document also gives detailed advice and recommendations on how the data should be submitted, with,

i) a description of the types of submission;
ii) what needs to be done when the submission is intended to expand the selection process criteria and precedes the development of a compound;
iii) or when the submission is intended to characterize a particular compound or,
iv) when the submission contains data supporting a general scientific discussion that is not necessarily related to the development of a compound.

The EMEA has also published a similar document "The Reflection Paper on Pharmacogenomic Samples, Testing and Data Handling" [29] which also covers most of the issues addressed in the FDA's draft guidance paper, but with less detail on the individual points.

Discussions between industry, academia and regulators focus on how far the regulation should extend into detail on the individual points and how much freedom should be left to the applicants. The evaluation of suitable, stable biological models used in "-omics" approaches, especially for preclinical models, and a minimum set of agreements and standardization on algorithms, analysis parameters, and documentation of data seem important, together with consistent monitoring of inter-laboratory variances.

The predictivity of toxicological models must be proven with a sufficient amount of reference datasets freely available to every stakeholder. It will be a major task for the future to build reference databases of expression profiles found to be predictive of certain classes of chemical substances, types of toxicity, disease, or therapeutic success. Collaborative efforts by international consortia are necessary, such as the projects the ILSI Health and Environmental Sciences Institute carried out with toxicologically-focused projects on substance- and organ-specific toxicity expression profiling [30], and are currently ongoing for biomarker qualification.

The pharmaceutical industry already has substantial experience with microarray experiments and some companies have substantial in-house databases [4, 31]. A thorough analysis and comparison of experiment and data quality and comparability of the different techniques used may result in a stock of datasets which may provide a substantial foundation of high quality data to start with.

The FDA's VGDS path and the EMEA's pharmacogenetics briefing meetings concept [9, 11] have introduced a novel approach for handling data submitted by the pharmaceutical industry. These concepts have created a forum for scientific discussion outside of regular review processes on early stage or exploratory pharmacogenomics data regarded as not yet ready for use in regulatory decision making.

Biomarker use and qualification

Transcriptional profiling analysis also has promising potential for identifying novel, predictive markers of toxicity. Development of robust safety markers would be of considerable value for early prediction of toxic liabilities of compounds, both in animal toxicology studies and in early, well-controlled clinical trials to monitor drug-induced pathologies when animal study findings present a cause for concern. However, the use of safety markers for these purposes requires extensive validation to establish appropriate sensitivity and specificity. For selected toxicities with broad impact but a lack of appropriate existing biomarkers (e.g. nephrotoxicity or non-genotoxic carcinogenesis), efforts for qualification of newly discovered biomarker candidates are currently underway, conducted by consortia such as the International Life Science Institute (ILSI, 2006) [http://www.ilsi.org/] and the Critical Path Institute (C-Path, 2006) [http://www.c-path.org].

As stated in the FDA Guidance for pharmacogenomic data submission [9], a known, valid biomarker is one that is broadly accepted in the scientific community and includes a consensus between scientists of the pharmaceutical industry, academia and regulatory agencies. However, for novel biomarkers with promising utility, sufficient knowledge may be available (e.g. in pharmaceutical companies) but not found in the public domain. It is therefore necessary to define a process agreeable to both industry and regulatory scientists for qualifying new biomarkers for drug development and regulatory decision-making.

A proposal for qualification of biomarkers for regulatory use has recently been published by scientists from the FDA [32, 33]. The C-Path Predictive Safety Testing Consortium (PSTC), a consortium of academia, industry and regulatory authorities, has started a process for qualification of novel biomarkers of nephrotoxicity [34]. This is of great interest, as currently used indicators of renal damage, such as blood urea nitrogen (BUN) or creatinine, are insensitive at identifying early damage and are found increased in blood only after severe damage to kidney functions. C-Path PSTC has submitted a package of data for qualifying novel biomarkers by the voluntary data submission path to both FDA and EMEA and a joint review process by regulatory scientists has been successfully completed [34]. Besides the attempt to qualify the specific biomarkers under review, it can be expected that this exercise will help establish a decision rendering process map for biomarker qualification acceptable to health authorities that would establish these biomarkers as acceptable regulatory tools.

Regulatory experience with pharmacogenomic data submissions

VGDSs [19, 35]

The FDA has received approximately 30 VGDSs within the two years since it has come into force. These VGDSs covered a broad range of therapeutic areas and genomic approaches for preclinical as well as all stages of clinical development. The focus of VGDSs so far, for example, concerned the pros and cons of candidate gene approaches *versus* whole genome SNP scans, expression profiling in peripheral blood cells, toxicogenomics, metabolomics and others. The overall response of industry delegates was positive, claiming they had gained greater understanding of how FDA analyzes and uses the data, learned more on how and what data to submit, and agreed that biological interpretation of the data is the core focus of review, and not individual lists of genes.

Voluntary pharmacogenetics briefing meetings [11]

Experience with and acceptance of voluntary briefing meetings at the EMEA is similar to that at the FDA. Around 20 meetings representing more than 30 case studies have been held so far. The cases also covered several therapeutic areas and all stages of preclinical and clinical development. It should also be noted that FDA and EMEA offer the opportunity of Joint Voluntary Briefing Meetings with both agencies. This opportunity has been used for four voluntary submissions so far.

With regular drug applications in the EU, the pharmaceutical industry has submitted toxicogenomics data with four submissions to the EMEA so far. In these cases toxicogenomics experiments provided supportive data for mechanistic evaluation of adverse effects seen in preclinical studies which had to be further investigated by the applicants after first review of submissions by regulatory bodies. The concerns identified were addressed by the applicants with classical additional toxicology studies and toxicogenomics approaches, which was also encouraged by regulators in such cases. Toxicogenomics data in both cases were biologically meaningful, supporting the proposed mechanisms of toxicity seen in animal studies.

Although toxicogenomics data have already been submitted, it is until now clear that these cases must be regarded as exceptions. As described by Foster 2007 [31] and Kasper 2005 [36] the main reasons for this reluctance to submit data are difficulties with the technique itself, significant variation in approaches of standards development and biomarker validation, and concern relative to regulatory approaches of data evaluation and requirements for reporting of transcriptional data. Uncertainties of possible mis- or over-interpretation of results and erroneous extrapolations from toxicogenomics data still exist. It should, however, be emphasized that regulatory agencies welcome submission of microarray data, as demonstrated by their various activities to encourage industry and academia to share their data and discuss interpretations with regulators.

References

1 Afshari CA, Nuwaysir EF, Barrett JC (1999) Application of complementary DNA microarray technology to carcinogen identification, toxicology and drug safety evaluation. *Cancer Res* 59: 4759–4760

2 Lovett RA (2000) Toxicologists brace for genomics revolution. *Science* 289: 536–537

3 Pollack A (2000) DNA chip may help usher in a new area of product testing. *New York Times* 28 Nov 2000

4 Lühe A, Suter L, Ruepp S, Singer T, Weiser T, Albertini S (2005) Toxicogenomics in the pharmaceutical industry: hollow promises or real benefit? *Mutat Res* 575: 102–115

5 Heidecker B, Hare JM (2007) The use of transcriptomic biomarkers for personalized medicine. *Heart Fail Rev* 12: 1–11

6 Fischer HP, Freiberg C (2007) Applications of transcriptional profiling in antibiotics discovery and development. *Prog Drug Res* 64: 21, 23–47
7 Lesko LJ, Woodcock J (2002) Pharmacogenomic-guided drug development: a regulatory perspective. *Pharmacogenomics* J 2: 20–24
8 FDA (2004) Challenge and opportunity on the critical path to new medical products (http://www.fda.gov/oc/initiatives/criticalpath/whitepaper.pdf)
9 FDA (2005) Guidance for Industry: Pharmacogenomic Data Submissions (http://www.fda.gov/cder/guidance/6400fnl.pdf)
10 FDA (2007) Guidance for industry and FDA staff: Pharmacogenetic tests and genetic tests for heritable markers (http://www.fda.gov/cdrh/oivd/guidance/1549.pdf)
11 EMEA (2006) Guideline on Pharmacogenetics Briefing Meetings (http://www.emea.europa.eu/pdfs/human/pharmacogenetics/2022704en.pdf)
12 FDA, EU, EMEA (2006) General principles: Processing Joint FDA EMEA Voluntary Genomic Data Submissions (VGDSs) within the framework of the Confidentiality Arrangement (http://www.emea.europa.eu/pdfs/human/pharmacogenetics/Guideline_on_Joint_VGDS_briefingmeetings.pdf)
13 ICH (2007) Final Draft Guideline E15: Definitions for genomic biomarkers, pharmacogenomics, pharmacogenetics, genomic data and sample coding catagories (http://www.ich.org/LOB/media/MEDIA3383.pdf)
14 Kola I, Landis J (2004) Can the pharmaceutical industry reduce attrition rates. *Nat Rev Drug Discov* 3: 711–715
15 FDA (2006) Table of valid genomic biomarkers in the context of approved drug labels (http://www.fda.gov/cder/genomics/genomic_biomarkers_table.htm) last updated Oct 27, 2006
16 Waters M, Yauk C (2007) Consensus recommendations to promote and advance predictive systems toxicology and toxicogenomics. *Environ Mol Mutagen* 48: 400–403
17 Lipshutz RJ, Morris D, Chee M, Hubbell E, Kozal MJ, Shah N, Shen N, Yang R, Fodor SP (1995) Using oligonucleotide probe arrays to access genetic diversity. *Biotechniques* 19: 442–447
18 Ramsay G (1998) DNA chips: state-of-the-art. *Nat Biotechnol* 16: 40–44
19 Frueh FW (2006) Impact of microarray data quality on genomic data submissions to the FDA. *Nat Biotechnol* 24: 1105–1107
20 Chan VSW, Theilade MD (2005) The use of toxicogenomic data in risk assessment: a regulatory perspective. *Clin Toxicology* 43: 121–126
21 Gruvberger-Saal SK, Cunliffe HE, Carr KM, Hedenfalk IA (2006) Microarrays in breast cancer research and clinical practice – the future lies ahead. *Endocr Rel Cancer* 13: 1017–1031
22 Surowiak P (2006) Prediction of the response to chemotherapy in ovarian cancers. *Folia Morphol* 65: 285–294
23 Achiron A, Gurevich M, Snir Y, Segal E, Mandel M (2007) Zinc-ion binding and cytokine activity regulation pathway predicts outcome in relapsing-remitting multiple sclerosis. *Clin Exp Immunol* 149: 235–242

24 Dietel M (2007) Predictive pathology of cytostatic drug resistance and new anti-cancer targets. *Recent Results Cancer Res* 176: 25–32

25 Brazma A, Hingamp P, Quackenbush J, Sherlock G, Spellman P, Stoeckert C, Aach J, Ansorge W, Ball CA, Causton HC et al (2001) Minimum information about a microarray experiment (MIAME)-toward standards for microarray data. *Nat Genet* 29: 365–371

26 The Microarray Quality Control Consortium (2006) *Nat Biotechnol* 24: 1103–1169

27 Pozhitkov AE, Tautz D, Noble PA (2007) Oligonucleotide microarrays: widely applied poorly understood. *Brief Funct Genomic Proteomic* 6: 141–148

28 FDA (2007) Draft Guidance for Industry – Pharmacogenomic data submissions – companion guidance (http://www.fda.gov/cder/guidance/7735dft.pdf)

29 EMEA (2007) Reflection paper on pharmacogenomic samples, testing and data handling (http://www.emea.europa.eu/pdfs/human/pharmacogenetics/20191406en.pdf)

30 Pennie W, Pettit, SD, Lord PG (2004) Toxicogenomics in risk assessment: An overview of an ILSI HESI collaborative research program. *Environ Health Perspect* 112: 417–419

31 Foster WR, Chen SJ, He A, Truong A, Bhaskaran V, Nelson DM, Dambach DM,Lehman-McKeeman LD, Car BD (2007) A retrospective analysis of Toxicogenomics in the safety assessment of drug candicates. *Toxicol Pathol* 35: 621–635

32 Goodsaid F, Frueh F (2006) Process map proposal for the validation of genomic biomarkers. *Pharmacogenomics* 7: 773–782

33 Goodsaid F, Frueh F (2007) Biomarker qualification pilot process at the US Food and Drug Administration. *AAPS J* 9: E105–E108.

34 EMEA (2008) First EMEA-FDA joint biomarker qualification process (http://www.emea.europa.eu/htms/human/mes/biomarkers.htm)

35 Orr MS, Goodsaid F, Amur S, Rudman A, Frueh FW (2007) The experience with voluntary genomic data submissions at the FDA and a vision for the future of the voluntary data submission program. *Clin Pharmacol Ther* 81: 294–297

36 Kasper P, Oliver G, Lima BS, Singer T, Tweats D (2005) Joint EFPIA/CHMP SWP Workshop: the Emerging Use of Omic Technologies for Regulatory Non-Clinical Safety Testing. *Pharmacogenomics* 6: 181–184

Outlook

Complementary microarray technologies

Bernhard Gerstmayer

Miltenyi Biotec GmbH, Friedrich-Ebert-Strasse 68, 51429 Bergisch Gladbach, Germany

Abstract

As outlined in the previous chapters of this book, the main microarray applications in inflammation rely on mRNA expression profiling based on either Oligo or cDNA microarray platforms. By virtue of measuring mRNA transcript levels, the activity of genes in inflammatory lesions can be analyzed. However, we have not focused solely on mRNA expression *via* microarray analysis over the last decade. Different disciplines from the genomics, glycomics, proteomics and metabolomics area have been combined in order to get an "all-inclusive" picture of the disease to be analysed. This combined approach has led to the creation of a new discipline named "systems biology". Novel microarray-based technologies have been developed that enable the analysis of "messenger" molecules other than mRNA. The focus of the current chapter is on those microarray platforms which either already play or are expected to play an important role in our understanding of the pathogenesis of inflammation. In addition, all described microarray platforms are meant to speed up drug discovery in future research and/or serve as prognostic or diagnostic tools in inflammatory diseases. This overview will concentrate on microarray platforms developed for promoter or CpG methylation as well as Chromatin Immunoprecipitation on chip analysis (ChIP on Chip), array comparative genomic hybridization (aCGH), carbohydrate microarrays, protein microarrays and microarrays for the detection and analysis of microRNAs.

DNA methylation microarrays and ChIP on Chip technology

Unlike microarrays for mRNA expression profiling, DNA methylation microarrays or ChIP on Chip (chromatin immunoprecipitation on chip) technologies concentrate on DNA levels as a read out.

DNA methylation or epigenetic methylation events are mechanisms that regulate gene expression without induction of changes within the DNA sequence itself. A common mode for epigenetic inheritance represents DNA modification of the base cytosine to 5-methylcytosine that occurs at 5'-CpG dinucleotides. Frequently, these modifications are observed in promoter regions of transcription factors that regulate e.g. developmental or other important regulatory processes, within the cell. If DNA hypermethylation or, in general, irregular methylation patterns occur in these

Microarrays in Inflammation, edited by Andreas Bosio and Bernhard Gerstmayer

promoter regions, this may lead to the development of automimmune diseases or cancer [1].

Using microarrays which harbor probes that target either specific promoter regions or regions where CpG islands are frequently observed, it is possible to detect methylation events on a genome-wide scale.

Several "methylated DNA" enrichment methods have been described. The two most commonly used approaches are illustrated in Figure 1. Briefly, cells or tissues are lysed, DNA is extracted and subsequently fragmented using either enzymatic digestion or ultrasonication. Enrichment of methylated DNA is performed either using a monoclonal antibody against 5-methylcytosine or *via* a more recently developed technique, named methylated-CpG island recovery assay (MIRA). The latter assay is based on the high affinity of the MBD2/MBD3L1 complex for methylated DNA and has been shown to detect cell type-dependent differences in DNA methylation [2]. Next, enriched methylated DNA is released from the antibody or protein complex under denaturing conditions, appropriate linkers are ligated to the DNA fragments followed by a PCR-based amplification step. Finally, input DNA and methylation enriched DNA are labeled with different dyes (mostly cyanine 3 and cyanine 5) using random priming. Both labeled samples are mixed and the competitive hybridization is performed on an appropriate microarray platform. Relative DNA methylation levels for each probe/CpG island are reflected in changes in cyanine 5/cyanine 3 ratios.

The main difference between DNA methylation microarray analysis *versus* ChIP on Chip analysis is that the latter combines chromatin immunoprecipitation (ChIP) with microarrays (chip) to analyze how regulatory proteins interact with the genome of living cells. These regulatory proteins – which also frequently encode transcription factors – bind to distinct chromosomal regions to control transcriptional activity or chromosomal replication. Identification of the target regions of these regulatory proteins is of crucial importance in gaining a better understanding of the involved molecular pathways and to help identify new target genes and therapeutics capable of modulating these pathways [3].

The overall workflow of ChIP on Chip analysis is very similar to the one described above for DNA methylation analysis. The main difference prior to DNA extraction, however, is that the cells to be analysed were treated with a fixative (mostly formaldehyde) in order to stabilize protein-DNA complexes. Lysis and fragmentation of DNA are similar for both platforms. After fragmentation, an immunoprecipitation step – ideally using a monoclonal (alternatively polyclonal) antibody – is performed. The protein-DNA complex is then hydrolyzed to reverse the cross-links within the released DNA fragments. Subsequently purified DNA fragments are amplified, labelled, hybridized and analysed similarly as described for the DNA methylation microarray workflow.

In summary, DNA methylation microarrays and ChIP on Chip technology provide important insights to vital processes such as e.g. proliferation, oncogenetic

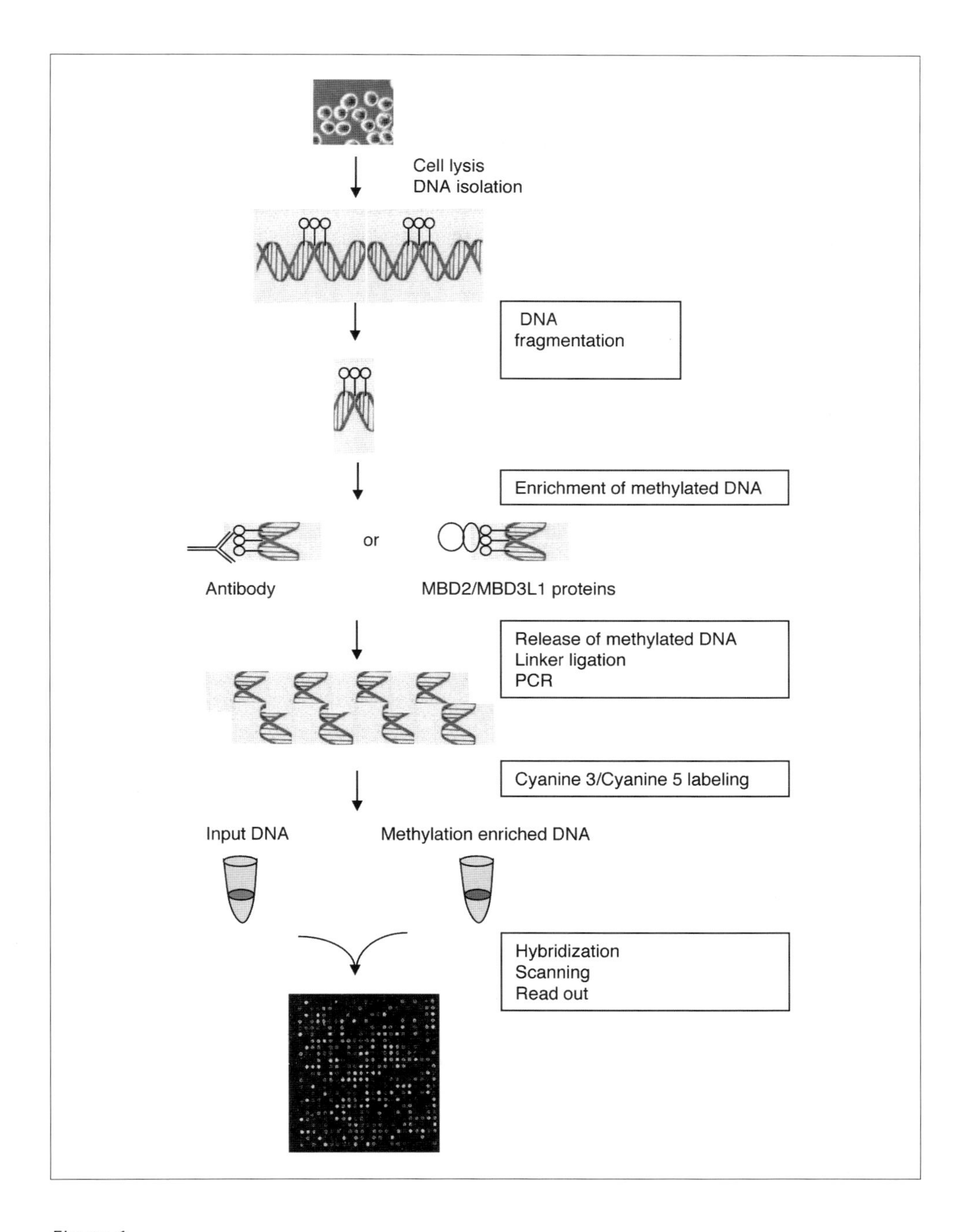

Figure 1.
Description of DNA methylation microarray workflow.

transformation or apoptosis. They will, therefore, speed up research on inflammatory diseases and are likely to contribute to drug discovery in this field.

Array comparative genomic hybridisation

Array comparative genomic hybridisation – like ChIP on Chip or methylation microarrays – has been developed to analyse genomic DNA levels in the samples of interest. aCGH provides a means to quantitatively measure DNA copy number variations and to map those regions to the genomic sequence. Identification of these chromosomal gains or losses can provide valuable information concerning origin, stage or status of genetic diseases and cancer [4].

Historically, detection of single genetic losses or deletions were performed using a cytogenetic technology named FISH (Fluorescence *In situ* Hybridization; [5, 6]). FISH makes use of sequence-specific fluorescently labeled probes that bind *via* Watson-Crick base pairing to their respective counterparts within the genome. Prior hybridization of the fluorescently labeled probe, a metaphase chromosomal preparation – usually on glass slides to firmly attach the DNA – has to be performed.

The reversal of this technique reflects the principle of the comparative genomic hybridization approach [7]. Here, hybridization is based on fluorescently labelled samples, e.g. tumor and reference DNA, to normal human chromosome preparations on glass slides. Several chromosomal aberrations can be detected at once using fluorescence microscopy and quantitative image analysis. However, because of the low genomic mapping resolution, only large chromosomal aberrations can be identified. Adaptation of this technique to an array platform resulted in the advent of aCGH [8]. The first successful aCGH protocols were based on PCR-amplified bacterial artificial chromosomes (BAC) spotted on glass slides [9]. Although at that time a technological revolution, PCR based BAC microarrays typically require maintenance, propagation, replication, and verification of large clone sets. To overcome these shortcomings, oligonucleotide microarray platforms are currently the technology of choice due to ease of handling and high reproducibility and resolution (in the kB range). If only single chromosomes or distinct regions within chromosomes are to be screened, usage of so-called tiling arrays achieves a resolution in the range of single bases.

Technically, the protocol is similar to the one described for DNA methylation analysis. Briefly, DNA is isolated from cells or tissues. In general, DNA from a healthy and diseased patient sample is then digested with restriction enzymes and each DNA is subsequently labeled with a distinct fluorescent dye. Both labeled samples are pooled and hybridized onto an appropriate microarray. After hybridization the two images for the respective dyes are visualized using a microarray scanner. The output always represents a comparison of genomes. It is noteworthy that aCGH allows only the detection of imbalanced copy-number variations (Fig. 2 A) but not aberrations where no net copy-number changes occur (Fig. 2 B).

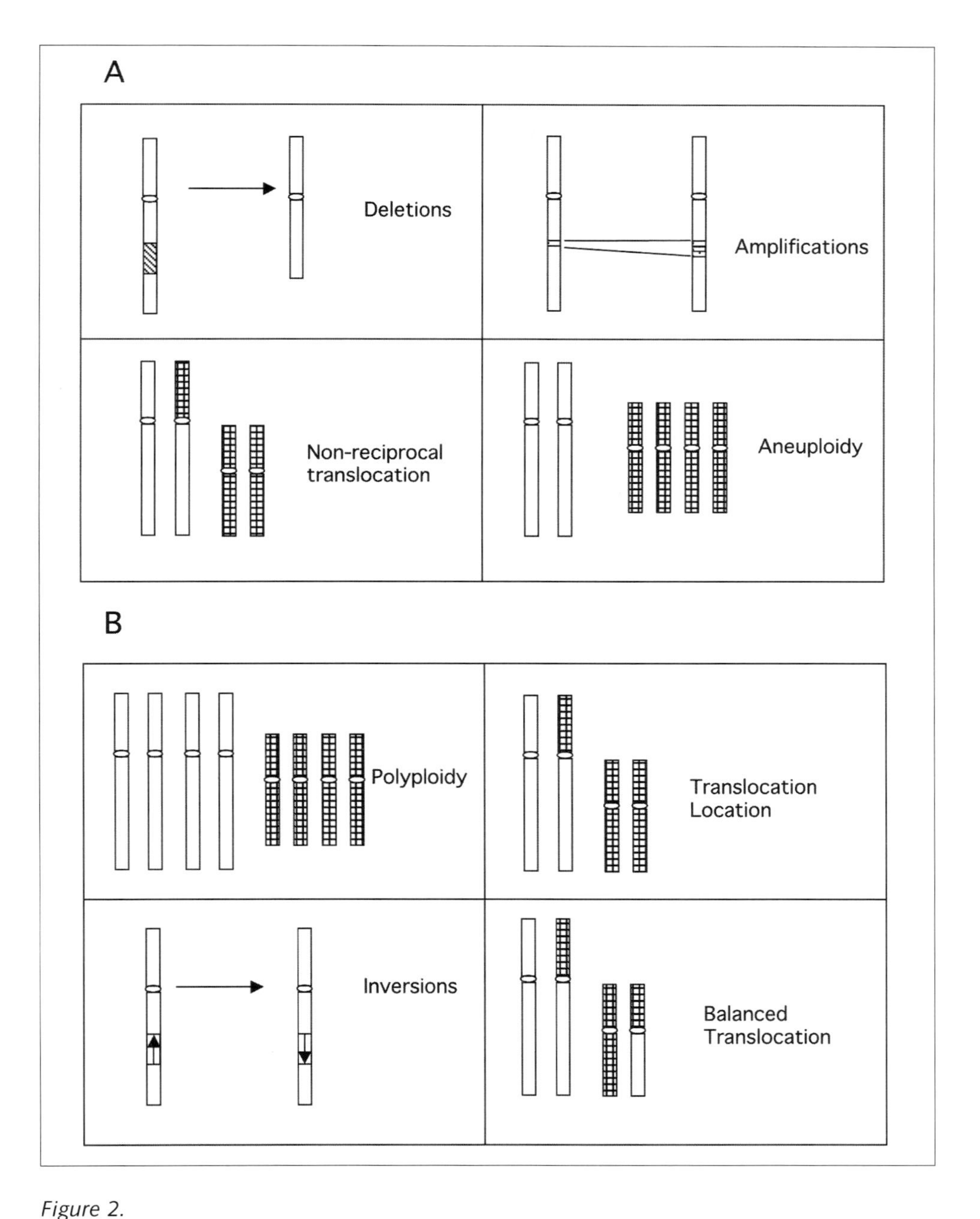

Figure 2.
A) Schematic representation of detectable imbalanced copy-number variations via *aCGH.*
B) Schematic representation of undetectable copy-number changes (no net copy-number variation).

Although the main application of aCGH currently focuses on cancer, where often large chromosomal regions are affected, it is very likely that the latest development of tiling aCGH arrays with ultrahigh resolution will also help identifying more subtle chromosomal changes in inflammatory diseases, especially in those diseases that appear to have a hereditary background.

Carbohydrate microarrays

In contrast to the above described "nucleotide microarray platforms" carbohydrate microarrays – as the name implies – measure carbohydrate expression levels. About half of all proteins synthesized within a eukaryotic cell carry carbohydrate or glycan chains (polymeric or complex carbohydrates). Carbohydrates play a number of crucial roles in a living organism. They not only serve as an energy currency e.g. as glycogen, but also have important functions in cell-cell recognition, host-pathogen interaction and protein integrity. Additionally, the rigid structure of glycan chains represents physical barriers that may prevent e.g. proteosomal attacks towards the cell. Therefore, dissecting the biological information contained in carbohydrate chains is of fundamental interest in the postgenomic area.

Unlike the central dogma, where the flow of information starts with DNA → transcribed into mRNA → translated into protein, there is no direct link between gene expression and glycan end product. Furthermore, standard molecular biology techniques such as PCR, cloning, and expression strategies cannot be employed for glycan research. Glycans are synthesized by different enzyme complexes and – after de novo synthesis – are transported and further modified in different compartments or organelles within the cell. Due to the challenges in immobilization of chemically and structurally complex glycans and the even more complex techniques required to synthesize them, development and progress of glycan microarrays have always lagged behind compared to DNA microarrays [10].

Several applications of carbohydrate microarrays have been described in the last six years. However, as discussed in the previous chapter for aCGH, currently the majority of carbohydrate microarray applications is dedicated to the field of oncology [11]. Nevertheless, this platform may also be employed to profile enzymatic activity, determine tissue specific-expression pattern, identify novel carbohydrate binding proteins or analyse lectin binding specificities in inflammatory diseases. One common characteristic pathogenic feature of inflammatory diseases like arthritis, asthma or diabetes represents leukocyte homing into affected tissues. Adhesion to endothelial cells lining the blood vessels and invasion of leucocytes into inflamed tissue are believed to represent one of the very first steps that initiate an inflammatory reaction. It has been shown that specific selectin ligands are absent in non-inflamed tissues but are expressed at high levels in endothelium of inflamed tissues. Deciphering the underlying carbohydrate-protein interactions using microarray approaches might be beneficial to target and interfere with pathogenic cell invasion [12, 13]. Furthermore, pharmacoglycomics can help to identify novel biomarkers or biomarker patterns that may provide valuable information for disease diagnosis or prediction [14].

In the year 2002, the first successful carbohydrate microarray platform was published [15]. Those microarrays – often referred to as Carbochips – were used to characterize carbohydrate–protein interactions. In another more recent report, novel glycan microarrays have been introduced that were manufactured and processed using similar tools as described for DNA microarrays [16]. There, standard robotic microarray printing technology, similar glass surface chemistry, scanner and imaging software were used to create a reliable and reproducible carbohydrate microarray platform.

It is beyond the scope of this chapter to discuss the different carbohydrate microarray platforms in detail; this has been done in reviews elsewhere [17-19]. However, one elegant and practical example of how carbohydrate–protein interactions can be analysed on a microarray platform will be highlighted in the following paragraph [16].

Briefly, a library of amino-functionalized glycans is spotted onto N-hydroxysuccinimide (NHS) activated glass slides to form covalent bonds. The microarray comprises about 200 synthetic and natural glycan sequences representing major glycan structures of glycoproteins and glycolipids. One goal within this study was to look for binding partners of individual glycan-binding proteins (GBP) or patient sera. Incubation of the GBPs or sera was performed under a cover slip for approximately 1 h at RT and bound complexes were detected using a fluorescent secondary antibody followed by visualization of the images *via* a microarray scanner. Although the focus of the described study was not on inflammatory diseases, analysis of patient sera from e.g. autoimmune patients or patients suffering from inflammatory diseases to identify specific glycan-protein interactions as targets for therapeutic intervention may be a very interesting topic for future research.

Protein microarrays

Unlike DNA microarrays that provide information about mRNA or gene expression levels but only sparse information on the functions of the proteins they encode, protein microarrays have been developed to close this gap. It is noteworthy, however, that proteins are completely different from RNA/DNA molecules with respect to their biochemical and biophysical properties. RNA/DNA, in contrast to proteins, represent molecules with a rather high stability in dissolved or dried form. In addition, RNA/DNA share – depending on their length and sequence – highly reproducible and predictable binding kinetics. The opposite is true of proteins. Each protein consists of a unique sequence made up of a repertoire of 20 different canonical amino acids. Due to this diverse repertoire, each protein has a unique polarity, a different size and three dimensional structure, a distinct isoelectric point and a different stability in solution. Since microarrays are generally packaged and shipped in a dry format, appropriate formulation of proteins on the microarray that preserves their function adds another layer of complexity.

Despite this complexity, a number of promising protein microarray platforms have been developed. Depending on the type of platform, analysis of protein-protein, protein-DNA, protein-drug, protein-phospholipid interactions can be performed [20, 21]. Furthermore, substrates of protein kinases have been identified using protein microarrays [22]. In the following paragraphs the most common platforms and their applications will be briefly summarized.

Basically, protein microarrays can be subdivided into four different categories: 1) analytical microarrays, 2) functional microarrays, 3) reverse phase microarrays or 4) self assembly protein microarrays.

The classic example of an analytical microarray is an antibody microarray (or less common, aptamer microarray). Here, a very distinct and uniform class of proteins – namely antibodies (or aptamers) – are spotted onto a microscopic slide. Antibody arrays are useful tools to measure binding affinities/specificities or protein expression levels of proteins in a complex mixture. Like DNA microarrays, antibody microarrays can be used to analyse differential expression of the target molecules of interest in e.g. diseased *versus* healthy tissues [23].

In contrast to analytical microarrays, functional protein microarrays consist of different spotted full length functional proteins or at least functional protein domains that allow the study of protein-protein, protein-DNA, protein-drug or protein-phospholipid interactions (reviewed in [24, 25]).

A prototypical example of the third protein microarray platform – reversed phase microarray (RPA) – represents spotting of cell or tissue lysates on nitrocellulose slides. Then, slides are incubated with an antibody of interest enabling the detection and quantification of the presence of the respective target protein in the lysate (reviewed in [26]).

Recently, a fourth promising protein platform has been described: Self assembly protein microarrays. There, the proteins are produced on the protein microarrays *in situ*. First, appropriate cDNA constructs that encode the proteins of interest are arrayed and immobilized on glass slides. Next, a mammalian reticulocyte lysate is added which simultaneously transcribes and translates the encoded proteins. The translated protein carries a tag which is encoded either at the 5' or 3' end of the cDNA sequence. Immediately after protein synthesis the tag is recognized by a capture protein (e.g. an antibody, which is co-immobilized on the slide surface) retaining the recombinant protein at the desired position on the microarray. A further refinement of this technology has recently been published: Nucleic acid programmable protein array (NAPPA, [27, 28]). This elegant "*in situ* protein synthesis approach" appears to minimize concerns raised about protein stability and integrity and it will be exciting to see how this technology evolves in the near future.

Different slide surfaces have been developed, depending on the platform type and whether the focus lies on integrity or functionality of the spotted proteins. As for DNA microarrays, amine-, aldehyde- and epoxy-derivatized glass surface chemistries can be used, allowing the proteins to covalently attach themselves to the surface in an oriented fashion [29]. In contrast, passive and randomized absorption of proteins can be achieved *via* coating the glass surface with e.g. nitrocellulose or poly-L-lysine [25]. As indicated earlier, production of protein microarrays using standard contact or non-contact printing devices has to be performed in a humidity-controlled environment to keep the proteins in their native state.

To label the samples to be analysed, similar techniques as described for DNA microarrays using e.g. radioisotope, affinity or fluorescent tags are applicable. Since the label itself might interfere with the sample's ability to bind to the target protein, appropriate labeling techniques have to be tested in advance. Recently, novel label-free detection methods (using e.g. microelectromechanical systems cantilevers) have been described that appear to be less invasive and may be superior to the standard techniques described above [30].

Taken together, the entire process of producing protein microarrays is a challenge. Several successful applications in inflammatory research, however, have been published recently, mainly based on protein profiling data derived from patient sera [31, 32]. Although, at the moment protein microarrays are primarily used as a research tool, once they become more robust and reproducible they are likely to get approval by health authorities and enter into clinical diagnostics.

microRNA microarrays

MicroRNAs (miRNAs) represent a distinct but highly informative, novel class of RNA molecules. As outlined in chapter 3.5 (Astrid Novosel and Arndt Borkhardt, miRNA patterns in haematopoietic malignancies) miRNAs are small, noncoding regulatory RNA molecules (21–24 nt) that control gene expression post-transcriptionally either by reducing stability and/or translation of fully or partially sequence-complementary target mRNAs.

Simultaneous analysis of miRNA expression represents the latest achievement that has been successfully translated into a microarray format [33–35]. In contrast to classic mRNA expression profiling, the establishment of a miRNA microarray platform represents a greater challenge with regard to probe selection, labeling and handling. First, due to the short length of miRNAs and their wide range of predicted melting temperatures *versus* their complementary DNA sequence, the hybridization temperature, the hybridization buffer as well as the hybridization kinetics have to be carefully adjusted. Second, although frequently a few miRNAs are expressed at very high levels in a certain tissue type (representing >50% of all miRNAs), the majority of miRNAs are mostly expressed at low copy numbers. In combination with their short sequence, which makes it difficult to incorporate more than one labelled molecule in their corresponding target sequence, detection of miRNAs represents another challenge.

Several miRNA isolation methods starting with tissue or cells are on the market. Currently the most commonly used method represents isolation of miRNA (together with total RNA!) *via* Trizol, a phenol/chloroform-based procedure. Most silica-based protocols do not allow the recovery of RNA molecules <100 nucleotides and are therefore not suitable for miRNA isolation.

Labeling of isolated miRNAs can be done in different ways. One option is to perform cDNA labeling during first strand synthesis using a fluorescence-tagged random octameric primer. Another option is to perform 3'-end-labeling using a ligase in the presence of fluorescently labeled nucleotides [36]. As for classic DNA microarrays, usually two different dyes for e.g. healthy *versus* diseased sample, are used.

Application of a universal reference that represents a defined pool of all synthetic miRNAs spotted on the array is an elegant way of reducing the amount of necessary sample if multiple samples need to be compared. By calculating the signal ratio of sample *versus* universal reference over the ratio of control *versus* universal reference, the resulting so-called re-ratio indirectly reflects the ratio of sample *versus* control.

Similar to gene expression profiling experiments, hybridization is usually performed overnight, ideally in an automated hybridization station to minimize artifacts due to handling and to increase reproducibility. Image analysis and read out can also be performed with standard equipment, although data analysis with respect to normalization does not appear to be a trivial task due to the generally rather low signal intensities of miRNAs.

In summary, miRNA microarrays represent a suitable tool to monitor miRNA expression in different disease types. In a recent publication miRNA expression profiling in peripheral blood cells of systemic lupus erythematosus (SLE) patients has been performed [37], indicating that miRNA microarrays will contribute to a better understanding of the pathogenic processes within autoimmune and inflammatory diseases.

References

1 Baylin, S.B., J.E. Ohm, Epigenetic gene silencing in cancer – a mechanism for early oncogenic pathway addiction? Nature reviews. *Cancer*, 2006. 6(2): 107–16

2 Rauch, T et al., MIRA-assisted microarray analysis, a new technology for the determi-

nation of DNA methylation patterns, identifies frequent methylation of homeodomain-containing genes in lung cancer cells. *Cancer Res*, 2006. 66(16): 7939–47

3 Boyer, L.A. et al., Core transcriptional regulatory circuitry in human embryonic stem cells. *Cell*, 2005. 122(6): 947–56

4 Barrett, M.T. et al., Comparative genomic hybridization using oligonucleotide microarrays and total genomic DNA. *Proc Nat Acad Sci USA*, 2004. 101(51): 17765–70. Epub 2004 Dec 10

5 Gall, J.G., M.L. Pardue, Formation and detection of RNA-DNA hybrid molecules in cytological preparations. *Proc Nat Acad Sci USA*, 1969. 63(2): 378–83

6 Pardue, M.L., J.G. Gall, Molecular hybridization of radioactive DNA to the DNA of cytological preparations. *Proc Nat Acad Sci USA*, 1969. 64(2): 600–4

7 Kallioniemi, A. et al., Comparative genomic hybridization for molecular cytogenetic analysis of solid tumors. *Science* (New York, N.Y.), 1992. 258(5083): 818–21

8 Pinkel, D. et al., High resolution analysis of DNA copy number variation using comparative genomic hybridization to microarrays. *Nature Genetics*, 1998. 20(2): 207–11

9 Greshock, J. et al., 1-Mb resolution array-based comparative genomic hybridization using a BAC clone set optimized for cancer gene analysis. *Genome Research*, 2004. 14(1): 179–87. Epub 2003 Dec 12

10 Drickamer, K., M.E. Taylor, Glycan arrays for functional glycomics. *Genome Biology*, 2002. 3(12): REVIEWS1034. Epub 2002 Nov 27

11 Kulig, P., J. Cichy, Acute phase mediator oncostatin M regulates affinity of alpha1-protease inhibitor for concanavalin A in hepatoma-derived but not lung-derived epithelial cells. *Cytokine*, 2005. 30(5): 269–74. Epub 2005 Mar 25

12 Renkonen, J. et al., Glycosylation might provide endothelial zip codes for organ-specific leukocyte traffic into inflammatory sites. *Am J Pathol*, 2002. 161(2): 543–50

13 Rosen, S.D, Endothelial ligands for L-selectin: from lymphocyte recirculation to allograft rejection. *Am J Pathol*, 1999. 155(4): 1013–20

14 Dube, D.H., C.R. Bertozzi, Glycans in cancer and inflammation – potential for therapeutics and diagnostics. Nature reviews. *Drug Discovery*, 2005. 4(6): 477–88

15 Wang, D. et al., Carbohydrate microarrays for the recognition of cross-reactive molecular markers of microbes and host cells. *Nature Biotechnology*, 2002. 20(3): 275–81

16 Blixt, O. et al., Printed covalent glycan array for ligand profiling of diverse glycan binding proteins. *Proc Nat Acad Sci USA*, 2004. 101(49): 17033–8. Epub 2004 Nov 24

17 Khan, I, D.V. Desai, A. Kumar, Carbochips: a new energy for old biobuilders. *J Biosci-Bioeng*, 2004. 98(5): 331–7

18 Miyamoto, S., Clinical applications of glycomic approaches for the detection of cancer and other diseases. *Curr Op Mol Ther*, 2006. 8(6): 507–13

19 Horlacher, T., P.H. Seeberger, The utility of carbohydrate microarrays in glycomics. *Omics: a Journal of Integrative Biology*, 2006. 10(4): 490–8

20 Hall, D.A. et al., Regulation of gene expression by a metabolic enzyme. *Science (New York, N.Y.)*, 2004. 306(5695): 482–4

21 Zhu, H., M. Snyder, Protein arrays and microarrays. *Curr Op Chem Biol*, 2001. 5(1): 40–5
22 Ptacek, J. et al., Global analysis of protein phosphorylation in yeast. *Nature*, 2005. 438(7068): 679–84
23 Sreekumar, A. et al., Profiling of cancer cells using protein microarrays: discovery of novel radiation-regulated proteins. *Cancer Res*, 2001. 61(20): 7585–93
24 Hall, D.A, J. Ptacek, M. Snyder, Protein microarray technology. *Mechanisms of Ageing and Development*, 2007. 128(1): 161–7. Epub 2006 Nov 28
25 Zhu, H., M. Snyder, Protein chip technology. *Curr Op Chem Biol*, 2003. 7(1): 55–63
26 Speer, R. et al., Reverse-phase protein microarrays for tissue-based analysis. *Curr Op Mol Ther*, 2005. 7(3): 240–5
27 Ramachandran, N. et al., Self-assembling protein microarrays. *Science (New York, N.Y.)*, 2004. 305(5680): 86–90
28 Ramachandran, N. et al., On-chip protein synthesis for making microarrays. *Methods Mol Biol* (Clifton, N.J.), 2006. 328: 1–14
29 Kusnezow, W., J.D. Hoheisel, Solid supports for microarray immunoassays. *Journal of Molecular Recognition: JMR*, 2003. 16(4): 165–76
30 Ramachandran, N. et al., Emerging tools for real-time label-free detection of interactions on functional protein microarrays. *FEBS*, 2005. 272(21): 5412–25
31 Pinto-Plata, V. et al., Profiling serum biomarkers in patients with COPD: associations with clinical parameters. *Thorax*, 2007. 62(7): 595–601. Epub 2007 Mar 13
32 Tabibiazar, R. et al., Proteomic profiles of serum inflammatory markers accurately predict atherosclerosis in mice. *Physiological Genomics*, 2006. 25(2): 194–202. Epub 2006 Jan 17
33 Landgraf, P. et al., A mammalian microRNA expression atlas based on small RNA library sequencing. *Cell*, 2007. 129(7): 1401–14
34 Liu, C.G. et al., Expression profiling of microRNA using oligo DNA arrays. *Methods (San Diego, Calif.)*, 2008. 44(1): 22–30
35 Castoldi, M. et al., miChip: an array-based method for microRNA expression profiling using locked nucleic acid capture probes. *Nature Protocols*, 2008. 3(2): 321–9
36 Wienholds, E. et al., MicroRNA expression in zebrafish embryonic development. *Science (New York, N.Y.)*, 2005. 309(5732): 310–1. Epub 2005 May 26
37 Dai, Y. et al., Microarray analysis of microRNA expression in peripheral blood cells of systemic lupus erythematosus patients. *Lupus*, 2007. 16(12): 939–46

Index

The PIR-Series
Progress in Inflammation Research

Homepage: http://www.birkhauser.ch

Up-to-date information on the latest developments in the pathology, mechanisms and therapy of inflammatory disease are provided in this monograph series. Areas covered include vascular responses, skin inflammation, pain, neuroinflammation, arthritis cartilage and bone, airways inflammation and asthma, allergy, cytokines and inflammatory mediators, cell signalling, and recent advances in drug therapy. Each volume is edited by acknowledged experts providing succinct overviews on specific topics intended to inform and explain. The series is of interest to academic and industrial biomedical researchers, drug development personnel and rheumatologists, allergists, pathologists, dermatologists and other clinicians requiring regular scientific updates.

Available volumes:

T Cells in Arthritis, P. Miossec, W. van den Berg, G. Firestein (Editors), 1998
Medicinal Fatty Acids, J. Kremer (Editor), 1998
Cytokines in Severe Sepsis and Septic Shock, H. Redl, G. Schlag (Editors), 1999
Cytokines and Pain, L. Watkins, S. Maier (Editors), 1999
Pain and Neurogenic Inflammation, S.D. Brain, P. Moore (Editors), 1999
Apoptosis and Inflammation, J.D. Winkler (Editor), 1999
Novel Inhibitors of Leukotrienes, G. Folco, B. Samuelsson, R.C. Murphy (Editors), 1999
Metalloproteinases as Targets for Anti-Inflammatory Drugs,
K.M.K. Bottomley, D. Bradshaw, J.S. Nixon (Editors), 1999
Gene Therapy in Inflammatory Diseases, C.H. Evans, P. Robbins (Editors), 2000
Cellular Mechanisms in Airways Inflammation, C. Page, K. Banner, D. Spina (Editors), 2000
Inflammatory and Infectious Basis of Atherosclerosis, J.L. Mehta (Editor), 2001
Neuroinflammatory Mechanisms in Alzheimer's Disease. Basic and Clinical Research,
J. Rogers (Editor), 2001
Inflammation and Stroke, G.Z. Feuerstein (Editor), 2001
NMDA Antagonists as Potential Analgesic Drugs,
D.J.S. Sirinathsinghji, R.G. Hill (Editors), 2002
Mechanisms and Mediators of Neuropathic pain, A.B. Malmberg, S.R. Chaplan (Editors), 2002
Bone Morphogenetic Proteins. From Laboratory to Clinical Practice,
S. Vukicevic, K.T. Sampath (Editors), 2002
The Hereditary Basis of Allergic Diseases, J. Holloway, S. Holgate (Editors), 2002
Inflammation and Cardiac Diseases, G.Z. Feuerstein, P. Libby, D.L. Mann (Editors), 2003
Mind over Matter – Regulation of Peripheral Inflammation by the CNS,
M. Schäfer, C. Stein (Editors), 2003
Heat Shock Proteins and Inflammation, W. van Eden (Editor), 2003
Pharmacotherapy of Gastrointestinal Inflammation, A. Guglietta (Editor), 2004
Arachidonate Remodeling and Inflammation, A.N. Fonteh, R.L. Wykle (Editors), 2004
Recent Advances in Pathophysiology of COPD, P.J. Barnes, T.T. Hansel (Editors), 2004
Cytokines and Joint Injury, W.B. van den Berg, P. Miossec (Editors), 2004

Cancer and Inflammation, D.W. Morgan, U. Forssmann, M.T. Nakada (Editors), 2004
Bone Morphogenetic Proteins: Bone Regeneration and Beyond, S. Vukicevic, K.T. Sampath (Editors), 2004
Antibiotics as Anti-Inflammatory and Immunomodulatory Agents, B.K. Rubin, J. Tamaoki (Editors), 2005
Antirheumatic Therapy: Actions and Outcomes, R.O. Day, D.E. Furst, P.L.C.M. van Riel, B. Bresnihan (Editors), 2005
Regulatory T-Cells in Inflammation, L. Taams, A.N. Akbar, M.H.M Wauben (Editors), 2005
Sodium Channels, Pain, and Analgesia, K. Coward, M. Baker (Editors), 2005
Turning up the Heat on Pain: TRPV1 Receptors in Pain and Inflammation, A.B Malmberg, K.R. Bley (Editors), 2005
The NPY Family of Peptides in Immune Disorders, Inflammation, Angiogenesis and Cancer, Z. Zukowska, G.Z. Feuerstein (Editors), 2005
Toll-like Receptors in Inflammation, L.A.J. O'Neill, E. Brint (Editors), 2005
Complement and Kidney Disease, P.F. Zipfel (Editor), 2006
Chemokine Biology – Basic Research and Clinical Application, Volume 1: Immunobiology of Chemokines, B. Moser, G.L. Letts, K. Neote (Editors), 2006
The Hereditary Basis of Rheumatic Diseases, R. Holmdahl (Editor), 2006
Lymphocyte Trafficking in Health and Disease, R. Badolato, S. Sozzani (Editors), 2006
In Vivo *Models of Inflammation, 2nd Edition, Volume I*, C.S. Stevenson, L.A. Marshall, D.W. Morgan (Editors), 2006
In Vivo *Models of Inflammation, 2nd Edition, Volume II*, C.S. Stevenson, L.A. Marshall, D.W. Morgan (Editors), 2006
Chemokine Biology – Basic Research and Clinical Application. Volume II: Pathophysiology of Chemokines, K. Neote, G.L. Letts, B. Moser (Editors), 2007
Adhesion Molecules: Function and Inhibition, K. Ley (Editor), 2007
The Immune Synapse as a Novel Target for Therapy, L. Graca (Editor), 2008
The Resolution of Inflammation, A.G. Rossi, D.A. Sawatzky (Editors), 2008
Bone Morphogenetic Proteins: From Local to Systemic Therapeutics, S. Vukicevic, K.T. Sampath (Editors), 2008
Angiogenesis in Inflammation: Mechanisms and Clinical Correlates, M.P. Seed, D.A. Walsh (Editors), 2008
Matrix Metalloproteinases in Tissue Remodelling and Inflammation, V. Lagente, E. Boichot (Editors), 2008